Jainism and Ecology

Publications of the Center for the Study of World Religions, Harvard Divinity School

General Editor: Lawrence E. Sullivan
Senior Editor: Kathryn Dodgson

Religions of the World and Ecology

Series Editors: Mary Evelyn Tucker and John Grim

Cambridge, Massachusetts

Jainism and Ecology

Nonviolence in the Web of Life

edited by

CHRISTOPHER KEY CHAPPLE

distributed by
Harvard University Press
for the
Center for the Study of World Religions
Harvard Divinity School

Printed in the United States of America

Grateful acknowledgment is made for permission to reprint the following:

Cartoon, *Jain Spirit*, no. 3 (March–May 2000).
L. M. Singhvi, *The Jain Declaration on Nature*, 1992.

Cataloging-in-Publication Data available from the Library of Congress

ISBN 0-945454-33-3 (hardcover)
ISBN 0-945454-34-1 (paperback)

Acknowledgments

The series of conferences on religions of the world and ecology took place from 1996 through 1998, with supervision at the Harvard University Center for the Study of World Religions by Don Kunkel and Malgorzata Radziszewska-Hedderick and with the assistance of Janey Bosch, Naomi Wilshire, and Lilli Leggio. Narges Moshiri, also at the Center, was indispensable in helping to arrange the first two conferences. A series of volumes developing the themes explored at the conferences is being published by the Center and distributed by Harvard University Press under the editorial direction of Kathryn Dodgson and with the skilled assistance of Eric Edstam.

These efforts have been generously supported by major funding from the V. Kann Rasmussen Foundation. The conference organizers appreciate also the support of the following institutions and individuals: Aga Khan Trust for Culture, Association of Shinto Shrines, Nathan Cummings Foundation, Dharam Hinduja Indic Research Center at Columbia University, Germeshausen Foundation, Harvard Buddhist Studies Forum, Harvard Divinity School Center for the Study of Values in Public Life, Jain Academic Foundation of North America, Laurance Rockefeller, Sacharuna Foundation, Theological Education to Meet the Environmental Challenge, and Winslow Foundation. The conferences were originally made possible by the Center for Respect of Life and Environment of the Humane Society of the United States, which continues to be a principal cosponsor. Bucknell University, also a cosponsor, has provided support in the form of leave time from teaching for conference coordinators Mary Evelyn Tucker and John

Grim as well as the invaluable administrative assistance of Stephanie Snyder. Her thoughtful attention to critical details is legendary. Then President William Adams of Bucknell University and then Vice-President for Academic Affairs Daniel Little have also granted travel funds for faculty and students to attend the conferences. Grateful acknowledgment is here made for the advice from key area specialists in shaping each conference and in editing the published volumes. Their generosity in time and talent has been indispensable at every step of the project. Throughout this process, the support, advice, and encouragement from Martin S. Kaplan has been invaluable.

I would like to thank several people who assisted with the preparation of this volume: Jin Kahng, Hope Blacker, Kathryn Dodgson, Mary Evelyn Tucker, John Grim, Louise Dobbs, Ericka Davis, Sylvere Valentin, Carol Turner, Stephanie Snyder, Barbara Michels, Michael Bennett, Nicole DePicotta, Tony Miranda, Casey Flaherty, and others. I am particularly grateful to Whitney Kelting and Don Davis for their suggestions for structuring the book. I would also like to express gratitude to the Jaina Academic Foundation of North America for its generous support of the conference on Jainism and Ecology held at the Center for the Study of World Religions, July 10–12, 1998. In particular, Dr. Sulekh Jain, Dr. Premchand Gada, and Dr. Manibhai Mehta were instrumental in providing support for this conference. The Boston Jaina Center extended its generous hospitality to the conference participants. I also wish to acknowledge the important contributions of Kim Skoog, B. C. Lodha, Cromwell Crawford, S. L. Gandhi, Samani Pratibha Prajna, and Samani Malli Prajna, whose presentations to the conference were deeply appreciated but are not included here in the published proceedings. Most especially, I want to thank the staff of the Center for the Study of World Religions, particularly Don Kunkel, Malgorzata Radziszewska-Hedderick, and Lawrence Sullivan for their support, and John Grim and Mary Evelyn Tucker, of Bucknell University, who served as the conference coordinators for the entire series of programs on the religions of the world and ecology.

Contents

Preface

LAWRENCE E. SULLIVAN

Religion distinguishes the human species from all others, just as human presence on earth distinguishes the ecology of our planet from other places in the known universe. Religious life and the earth's ecology are inextricably linked, organically related.

Human belief and practice mark the earth. One can hardly think of a natural system that has not been considerably altered, for better or worse, by human culture. "Nor is this the work of the industrial centuries," observes Simon Schama. "It is coeval with the entirety of our social existence. And it is this irreversibly modified world, from the polar caps to the equatorial forests, that is all the nature we have" (*Landscape and Memory* [New York: Vintage Books, 1996], 7). In Schama's examination even landscapes that appear to be most free of human culture turn out, on closer inspection, to be its product.

Human beliefs about the nature of ecology are the distinctive contribution of our species to the ecology itself. Religious beliefs—especially those concerning the nature of powers that create and animate—become an effective part of ecological systems. They attract the power of will and channel the forces of labor toward purposive transformations. Religious rituals model relations with material life and transmit habits of practice and attitudes of mind to succeeding generations.

This is not simply to say that religious thoughts occasionally touch the world and leave traces that accumulate over time. The matter is the other way around. From the point of view of environmental studies, religious worldviews propel communities into the world with fundamental predispositions toward it because such religious worldviews are primordial, all-encompassing, and unique. They are *primordial* because they probe behind secondary appearances and stray thoughts to rivet human attention on realities of the first order: life at its source, creativity in its fullest manifestation, death and destruction at their origin, renewal and salvation in their germ. The revelation of first things is compelling and moves communities to take creative action. Primordial ideas are prime movers.

Religious worldviews are *all-encompassing* because they fully absorb the natural world within them. They provide human beings both a view of the whole and at the same time a penetrating image of their own ironic position as the beings in the cosmos who possess the capacity for symbolic thought: the part that contains the whole—or at least a picture of the whole—within itself. As all-encompassing, therefore, religious ideas do not just contend with other ideas as equals; they frame the mind-set within which all sorts of ideas commingle in a cosmology. For this reason, their role in ecology must be better understood.

Religious worldviews are *unique* because they draw the world of nature into a wholly other kind of universe, one that appears only in the religious imagination. From the point of view of environmental studies, the risk of such religious views, on the one hand, is of disinterest in or disregard for the natural world. On the other hand, only in the religious world can nature be compared and contrasted to other kinds of being—the supernatural world or forms of power not always fully manifest in nature. Only then can nature be revealed as distinctive, set in a new light startlingly different from its own. That is to say, only religious perspectives enable human beings to evaluate the world of nature in terms distinct from all else. In this same step toward intelligibility, the natural world is evaluated in terms consonant with human beings' own distinctive (religious and imaginative) nature in the world, thus grounding a self-conscious relationship and a role with limits and responsibilities.

In the struggle to sustain the earth's environment as viable for future generations, environmental studies has thus far left the role of religion unprobed. This contrasts starkly with the emphasis given, for example, the role of science and technology in threatening or sustaining the ecology. Ignorance of religion prevents environmental studies from achieving its goals, however, for though science and technology share many important features of human culture with religion, they leave unexplored essential wellsprings of human motivation and concern that shape the world as we know it. No understanding of the environment is adequate without a grasp of the religious life that constitutes the human societies which saturate the natural environment.

A great deal of what we know about the religions of the world is new knowledge. As is the case for geology and astronomy, so too for religious studies: many new discoveries about the nature and function of religion are, in fact, clearer understandings of events and processes that began to unfold long ago. Much of what we are learning now about the religions of the world was previously not known outside of a circle of adepts. From the ancient history of traditions and from the ongoing creativity of the world's contemporary religions we are opening a treasury of motives, disciplines, and awarenesses.

A geology of the religious spirit of humankind can well serve our need to relate fruitfully to the earth and its myriad life-forms. Changing our habits of consumption and patterns of distribution, reevaluating modes of production, and reestablishing a strong sense of solidarity with the matrix of material life—these achievements will arrive along with spiritual modulations that unveil attractive new images of well-being and prosperity, respecting the limits of life in a sustainable world while revering life at its sources. Remarkable religious views are presented in this series—from the nature mysticism of Bashō in Japan or Saint Francis in Italy to the ecstatic physiologies and embryologies of shamanic healers, Taoist meditators, and Vedic practitioners; from indigenous people's ritual responses to projects funded by the World Bank, to religiously grounded criticisms of hazardous waste sites, deforestation, and environmental racism.

The power to modify the world is both frightening and fascinating and has been subjected to reflection, particularly religious reflection, from time immemorial to the present day. We will understand ecology better when we understand the religions that form the rich soil of memory and practice, belief and relationships where life on earth is rooted. Knowledge of these views will help us reappraise our ways and reorient ourselves toward the sources and resources of life.

This volume is one in a series that addresses the critical gap in our contemporary understanding of religion and ecology. The series results from research conducted at the Harvard University Center for the Study of World Religions over a three-year period. I wish especially to acknowledge President Neil L. Rudenstine of Harvard University for his leadership in instituting the environmental initiative at Harvard and thank him for his warm encouragement and characteristic support of our program. Mary Evelyn Tucker and John Grim of Bucknell University coordinated the research, involving the direct participation of some six hundred scholars, religious leaders, and environmental specialists brought to Harvard from around the world during the period of research and inquiry. Professors Tucker and Grim have brought great vision and energy to this enormous project, as has their team of conference convenors. The commitment and advice of Martin S. Kaplan of Hale and Dorr have been of great value. Our goals have been achieved for this research and publication program because of the extraordinary dedication and talents of Center for the Study of World Religions staff members Don Kunkel, Malgorzata Radziszewska-Hedderick, Kathryn Dodgson, Janey Bosch, Naomi Wilshire, Lilli Leggio, and Eric Edstam and with the unstinting help of Stephanie Snyder of Bucknell. To these individuals, and to all the sponsors and participants whose efforts made this series possible, go deepest thanks and appreciation.

Series Foreword

MARY EVELYN TUCKER and JOHN GRIM

The Nature of the Environmental Crisis

Ours is a period when the human community is in search of new and sustaining relationships to the earth amidst an environmental crisis that threatens the very existence of all life-forms on the planet. While the particular causes and solutions of this crisis are being debated by scientists, economists, and policymakers, the facts of widespread destruction are causing alarm in many quarters. Indeed, from some perspectives the future of human life itself appears threatened. As Daniel Maguire has succinctly observed, "If current trends continue, we will not."[1] Thomas Berry, the former director of the Riverdale Center for Religious Research, has also raised the stark question, "Is the human a viable species on an endangered planet?"

From resource depletion and species extinction to pollution overload and toxic surplus, the planet is struggling against unprecedented assaults. This is aggravated by population explosion, industrial growth, technological manipulation, and military proliferation heretofore unknown by the human community. From many accounts the basic elements which sustain life—sufficient water, clean air, and arable land—are at risk. The challenges are formidable and well documented. The solutions, however, are more elusive and complex. Clearly, this crisis has economic, political, and social dimensions which require more detailed analysis than we can provide here. Suffice it to say, however, as did the *Global 2000 Report*: ". . .once such global environmental problems are in motion they are difficult to reverse. In fact few if any of the problems addressed in the *Global 2000*

Report are amenable to quick technological or policy fixes; rather, they are inextricably mixed with the world's most perplexing social and economic problems."[2]

Peter Raven, the director of the Missouri Botanical Garden, wrote in a paper titled "We Are Killing Our World" with a similar sense of urgency regarding the magnitude of the environmental crisis: "The world that provides our evolutionary and ecological context is in serious trouble, trouble of a kind that demands our urgent attention. By formulating adequate plans for dealing with these large-scale problems, we will be laying the foundation for peace and prosperity in the future; by ignoring them, drifting passively while attending to what may seem more urgent, personal priorities, we are courting disaster."

Rethinking Worldviews and Ethics

For many people an environmental crisis of this complexity and scope is not only the result of certain economic, political, and social factors. It is also a moral and spiritual crisis which, in order to be addressed, will require broader philosophical and religious understandings of ourselves as creatures of nature, embedded in life cycles and dependent on ecosystems. Religions, thus, need to be reexamined in light of the current environmental crisis. This is because religions help to shape our attitudes toward nature in both conscious and unconscious ways. Religions provide basic interpretive stories of who we are, what nature is, where we have come from, and where we are going. This comprises a worldview of a society. Religions also suggest how we should treat other humans and how we should relate to nature. These values make up the ethical orientation of a society. Religions thus generate worldviews and ethics which underlie fundamental attitudes and values of different cultures and societies. As the historian Lynn White observed, "What people do about their ecology depends on what they think about themselves in relation to things around them. Human ecology is deeply conditioned by beliefs about our nature and destiny—that is, by religion."[3]

In trying to reorient ourselves in relation to the earth, it has become apparent that we have lost our appreciation for the intricate nature of matter and materiality. Our feeling of alienation in the modern period has extended beyond the human community and its patterns of material exchanges to our interaction with nature itself. Especially in technologically sophisticated urban societies, we have become removed from the recognition of our dependence on nature. We no longer know who we are as earthlings; we no longer see the earth as sacred.

Thomas Berry suggests that we have become autistic in our interactions with the natural world. In other words, we are unable to value the life and beauty of nature because we are locked in our own egocentric perspectives and shortsighted needs. He suggests that we need a new cosmology, cultural coding, and motivating energy to overcome this deprivation.[4] He observes that the magnitude of destructive industrial processes is so great that we must initiate a radical rethinking of the myth of progress and of humanity's role in the evolutionary process. Indeed, he speaks of evolution as a new story of the universe, namely, as a vast cosmological perspective that will resituate human meaning and direction in the context of four and a half billion years of earth history.[5]

For Berry and for many others an important component of the current environmental crisis is spiritual and ethical. It is here that the religions of the world may have a role to play in cooperation with other individuals, institutions, and initiatives that have been engaged with environmental issues for a considerable period of time. Despite their lateness in addressing the crisis, religions are beginning to respond in remarkably creative ways. They are not only rethinking their theologies but are also reorienting their sustainable practices and long-term environmental commitments. In so doing, the very nature of religion and of ethics is being challenged and changed. This is true because the reexamination of other worldviews created by religious beliefs and practices may be critical to our recovery of sufficiently comprehensive cosmologies, broad conceptual frameworks, and effective environmental ethics for the twenty-first century.

While in the past none of the religions of the world have had to face an environmental crisis such as we are now confronting, they remain key instruments in shaping attitudes toward nature. The unintended consequences of the modern industrial drive for unlimited economic growth and resource development have led us to an impasse regarding the survival of many life-forms and appropriate management of varied ecosystems. The religious traditions may indeed be critical in helping to reimagine the viable conditions and long-range strategies for fostering mutually enhancing human-earth relations.[6]

Indeed, as E. N. Anderson has documented with impressive detail, "All traditional societies that have succeeded in managing resources well, over time, have done it in part through religious or ritual representation of resource management."[7]

It is in this context that a series of conferences and publications exploring the various religions of the world and their relation to ecology was initiated by the Center for the Study of World Religions at Harvard. Coordinated by Mary Evelyn Tucker and John Grim, the conferences involved some six hundred scholars, graduate students, religious leaders, and environmental activists over a period of three years. The collaborative nature of the project is intentional. Such collaboration maximizes the opportunity for dialogical reflection on this issue of enormous complexity and accentuates the diversity of local manifestations of ecologically sustainable alternatives.

This series is intended to serve as initial explorations of the emerging field of religion and ecology while pointing toward areas for further research. We are not unaware of the difficulties of engaging in such a task, yet we have been encouraged by the enthusiastic response to the conferences within the academic community, by the larger interest they have generated beyond academia, and by the probing examinations gathered in the volumes. We trust that this series and these volumes will be useful not only for scholars of religion but also for those shaping seminary education and institutional religious practices, as well as for those involved in public policy on environmental issues.

We see such conferences and publications as expanding the growing dialogue regarding the role of the world's religions as moral forces in stemming the environmental crisis. While, clearly, there are major methodological issues involved in utilizing traditional philosophical and religious ideas for contemporary concerns, there are also compelling reasons to support such efforts, however modest they may be. The world's religions in all their complexity and variety remain one of the principal resources for symbolic ideas, spiritual inspiration, and ethical principles. Indeed, despite their limitations, historically they have provided comprehensive cosmologies for interpretive direction, moral foundations for social cohesion, spiritual guidance for cultural expression, and ritual celebrations for meaningful life. In our search for more comprehensive ecological worldviews and more effective environmental ethics, it is inevitable that we will draw from the symbolic and conceptual resources of the religious traditions of the world. The effort to do this is not without precedent or problems, some of which will be signaled below. With this volume and with this series we hope the field of reflection and discussion regarding religion and ecology will begin to broaden, deepen, and complexify.

Qualifications and Goals

The Problems and Promise of Religions

These volumes, then, are built on the premise that the religions of the world may be instrumental in addressing the moral dilemmas created by the environmental crisis. At the same time we recognize the limitations of such efforts on the part of religions. We also acknowledge that the complexity of the problem requires interlocking approaches from such fields as science, economics, politics, health, and public policy. As the human community struggles to formulate different attitudes toward nature and to articulate broader conceptions of ethics embracing species and ecosystems, religions may thus be a necessary, though only contributing, part of this multidisciplinary approach.

It is becoming increasingly evident that abundant scientific knowledge of the crisis is available and numerous political and economic statements have been formulated. Yet we seem to lack the political, economic, and scientific leadership to make necessary changes. Moreover, what is still lacking is the religious commitment, moral imagination, and ethical engagement to transform the environmental crisis from an issue on paper to one of effective policy, from rhetoric in print to realism in action. Why, nearly fifty years after Fairfield Osborne's warning in *Our Plundered Planet* and more than thirty years since Rachel Carson's *Silent Spring,* are we still wondering, is it too late?[8]

It is important to ask where the religions have been on these issues and why they themselves have been so late in their involvement. Have issues of personal salvation superseded all others? Have divine-human relations been primary? Have anthropocentric ethics been all-consuming? Has the material world of nature been devalued by religion? Does the search for otherworldly rewards override commitment to this world? Did the religions simply surrender their natural theologies and concerns with exploring purpose in nature to positivistic scientific cosmologies? In beginning to address these questions, we still have not exhausted all the reasons for religions' lack of attention to the environmental crisis. The reasons may not be readily apparent, but clearly they require further exploration and explanation.

In discussing the involvement of religions in this issue, it is also appropriate to acknowledge the dark side of religion in both its institutional expressions and dogmatic forms. In addition to their oversight with regard to the environment, religions have been the source of enormous manipulation of power in fostering wars, in ignoring racial and social injustice, and in promoting unequal gender relations, to name only a few abuses. One does not want to underplay this shadow side or to claim too much for religions' potential for ethical persuasiveness. The problems are too vast and complex for unqualified optimism. Yet there is a growing consensus that religions may now have a significant role to play, just as in the past they have sustained individuals and cultures in the face of internal and external threats.

A final caveat is the inevitable gap that arises between theories and practices in religions. As has been noted, even societies with religious traditions which appear sympathetic to the environment have in the past often misused resources. While it is clear that religions may have some disjunction between the ideal and the real, this should not lessen our endeavor to identify resources from within the world's religions for a more ecologically sound cosmology and environmentally supportive ethics. This disjunction of theory and practice is present within all philosophies and religions and is frequently the source of disillusionment, skepticism, and cynicism. A more realistic observation might be made, however, that this disjunction should not automatically invalidate the complex worldviews and rich cosmologies embedded in traditional religions. Rather, it is our task to explore these conceptual resources so as to broaden and expand our own perspectives in challenging and fruitful ways.

In summary, we recognize that religions have elements which are both prophetic and transformative as well as conservative and constraining. These elements are continually in tension, a condition which creates the great variety of thought and interpretation within religious traditions. To recognize these various tensions and limits, however, is not to lessen the urgency of the overall goals of this project. Rather, it is to circumscribe our efforts with healthy skepticism, cautious optimism, and modest ambitions. It is to suggest that this is a beginning in a new field of study which will affect both religion and ecology. On the one hand, this process of reflection will inevitably change how religions conceive of their own roles, missions, and identities, for such reflections demand a new sense of the sacred as not divorced from the earth itself. On the other hand, environmental studies can recognize that religions have helped to shape attitudes toward nature. Thus, as religions themselves evolve they may be indispensable in fostering a more expansive appreciation for the complexity and beauty of the natural world. At the same time as religions foster awe and reverence for nature, they may provide the transforming energies for ethical practices to protect endangered ecosystems, threatened species, and diminishing resources.

Methodological Concerns

It is important to acknowledge that there are, inevitably, challenging methodological issues involved in such a project as we are undertaking in this emerging field of religion and ecology.[9] Some of the key interpretive challenges we face in this project concern issues of time, place, space, and positionality. With regard to time, it is necessary to recognize the vast historical complexity of each religious tradition, which cannot be easily condensed in these conferences or volumes. With respect to place, we need to signal the diverse cultural contexts in which these religions have developed. With regard to space, we recognize the varied frameworks of institutions and traditions in which these religions unfold. Finally, with respect to positionality, we acknowledge our own historical situatedness at the end of the twentieth century with distinctive contemporary concerns.

Not only is each religious tradition historically complex and culturally diverse, but its beliefs, scriptures, and institutions have themselves been subject to vast commentaries and revisions over time. Thus, we recognize the radical diversity that exists within and among religious traditions which cannot be encompassed in any single volume. We acknowledge also that distortions may arise as we examine earlier historical traditions in light of contemporary issues.

Nonetheless, the environmental ethics philosopher J. Baird Callicott has suggested that scholars and others "mine the conceptual resources" of the religious traditions as a means of creating a more inclusive global environmental ethics.[10] As Callicott himself notes, however, the notion of "mining" is problematic, for it conjures up images of exploitation which may cause apprehension among certain religious communities, especially those of indigenous peoples. Moreover, we cannot simply expect to borrow or adopt ideas and place them from one tradition directly into another. Even efforts to formulate global environmental ethics need to be sensitive to cultural particularity and diversity. We do not aim at creating a simple bricolage or bland fusion of perspectives. Rather, these conferences and volumes are an attempt to display before us a multiperspectival cross section of the symbolic richness regarding attitudes toward nature within the religions of the world. To do so will help to reveal certain commonalities among traditions, as well as limitations within traditions, as they begin to converge around this challenge presented by the environmental crisis.

We need to identify our concerns, then, as embedded in the constraints of our own perspectival limits at the same time as we seek common ground. In describing various attitudes toward nature historically, we are aiming at *critical understanding* of the complexity, contexts, and frameworks in which these religions articulate such views. In addition, we are striving for *empathetic appreciation* for the traditions without idealizing their ecological potential or ignoring their environmental oversights. Finally, we are aiming at the *creative revisioning* of mutually enhancing human-earth relations. This revisioning may be assisted by highlighting the multiperspectival attitudes toward nature which these traditions disclose. The prismatic effect of examining such attitudes and relationships may provide some necessary clarification and symbolic resources for reimagining our own situation and shared concerns at the end of the twentieth century. It will also be sharpened by identifying the multilayered symbol systems in world religions which have traditionally oriented humans in establishing relational resonances between the microcosm of the self and the macrocosm of the social and natural orders. In short, religious traditions may help to supply both creative resources of symbols, rituals, and texts as well as inspiring visions for reimagining ourselves as part of, not apart from, the natural world.

Aims

The methodological issues outlined above were implied in the overall goals of the conferences, which were described as follows:

1. To identify and evaluate the *distinctive ecological attitudes,* values, and practices of diverse religious traditions, making clear their links to intellectual, political, and other resources associated with these distinctive traditions.

2. To describe and analyze the *commonalities* that exist within and among religious traditions with respect to ecology.

3. To identify the *minimum common ground* on which to base constructive understanding, motivating discussion, and concerted action in diverse locations across the globe; and to highlight the specific religious resources that comprise such fertile ecological ground: within scripture, ritual, myth, symbol, cosmology, sacrament, and so on.

4. To articulate in clear and moving terms *a desirable mode of human presence with the earth;* in short, to highlight means of respecting and valuing nature, to note what has already been actualized, and to indicate how best to achieve what is desirable beyond these examples.

5. To outline the most significant areas, with regard to religion and ecology, in need of *further study;* to enumerate questions of highest priority within those areas and propose possible approaches to use in addressing them.

In this series, then, we do not intend to obliterate difference or ignore diversity. The aim is to celebrate plurality by raising to conscious awareness multiple perspectives regarding nature and human-earth relations as articulated in the religions of the world. The spectrum of cosmologies, myths, symbols, and rituals within the religious traditions will be instructive in resituating us within the rhythms and limits of nature.

We are not looking for a unified worldview or a single global ethic. We are, however, deeply sympathetic with the efforts toward formulating a global ethic made by individuals, such as the theologian Hans Küng or the environmental philosopher J. Baird Callicott, and groups, such as Global Education Associates and United Religions. A minimum content of environmental ethics needs to be seriously considered. We are, then, keenly interested in the contribution this series might make to discussions of environmental policy in national and international arenas. Important intersections may be made with work in the field of development ethics.[11] In addition, the findings of the conferences have bearing on the ethical formulation of the Earth Charter that is to be presented to the United Nations for adoption within the next few years. Thus, we are seeking both the grounds for common concern and the constructive conceptual basis for rethinking our current situation of estrangement from the earth. In so doing we will be able to reconceive a means of creating the basis not just for sustainable development, but also for sustainable life on the planet.

As scientist Brian Swimme has suggested, we are currently making macrophase changes to the life systems of the planet with microphase wisdom. Clearly, we need to expand and deepen the wisdom base for human intervention with nature and other humans. This is particularly true as issues of genetic alteration of natural processes are already available and in use. If religions have traditionally concentrated on divine-human and human-human relations, the challenge is that they now explore more fully divine-human-earth relations. Without such further exploration, adequate environmental ethics may not emerge in a comprehensive context.

Resources: Environmental Ethics Found in the World's Religions

For many people, when challenges such as the environmental crisis are raised in relation to religion in the contemporary world, there frequently arises a sense of loss or a nostalgia for earlier, seemingly less complicated eras when the constant questioning of religious beliefs and practices was not so apparent. This is, no doubt, something of a reified reading of history. There is, however, a decidedly anxious tone to the questioning and soul-searching that appears to haunt many contemporary religious groups as they seek to find their particular role in the midst of rapid technological change and dominant secular values.

One of the greatest challenges, however, to contemporary religions remains how to respond to the environmental crisis, which many believe has been perpetuated because of the enormous inroads made by unrestrained materialism, secularization, and industrialization in contemporary societies, especially those societies arising in or influenced by the modern West. Indeed, some suggest that the very division of religion from secular life may be a major cause of the crisis.

Others, such as the medieval historian Lynn White, have cited religion's negative role in the crisis. White has suggested that the emphasis in Judaism and Christianity on the transcendence of God above nature and the dominion of humans over nature has led to a devaluing of the natural world and a subsequent destruction of its resources for utilitarian ends.[12] While the particulars of this argument have been vehemently debated, it is increasingly clear that the environmental crisis and its perpetuation due to industrialization, secularization, and ethical indifference present a serious challenge to the world's religions. This is especially true because many of these religions have traditionally been concerned with the path of personal salvation, which frequently emphasized otherworldly goals and rejected this world as corrupting. Thus, as we have noted, how to adapt religious teachings to this task of revaluing nature so as to prevent its destruction marks a significant new phase in religious thought. Indeed, as Thomas Berry has so aptly pointed out, what is necessary is a comprehensive reevaluation of human-earth relations if the human is to continue as a viable species on an increasingly degraded planet. This will require, in addition to major economic and political changes, examining worldviews and ethics among the world's religions that differ from those that have captured the imagination of contemporary industrialized societies which regard nature primarily as a commodity to be utilized. It should be noted that when we are searching for effective resources for formulating environmental ethics, each of the religious traditions have both positive and negative features.

For the most part, the worldviews associated with the Western Abrahamic traditions of Judaism, Christianity, and Islam have created a dominantly human-focused morality. Because these worldviews are largely anthropocentric, nature is viewed as being of secondary importance. This is reinforced by a strong sense of the transcendence of God above nature. On the other hand, there are rich resources for rethinking views of nature in the covenantal tradition of the Hebrew Bible, in sacramental theology, in incarnational Christology, and in the vice-regency (*khalifa Allah*) concept of the Qur'an. The covenantal tradition draws on the legal agreements of biblical thought which are extended to all of creation. Sacramental theology in Christianity underscores the sacred dimension of material reality, especially for ritual purposes.[13] Incarnational Christology proposes that because God became flesh in the person of Christ, the entire natural order can be viewed as sacred. The concept of humans as vice-regents of Allah on earth suggests that humans have particular privileges, responsibilities, and obligations to creation.[14]

In Hinduism, although there is a significant emphasis on performing one's *dharma,* or duty, in the world, there is also a strong pull toward *mokśa,* or liberation, from the world of suffering, or *saṃsāra*. To heal this kind of suffering and alienation through spiritual discipline and meditation, one turns away from the world (*prakṛti*) to a timeless world of spirit (*puruṣa*). Yet at the same time there are numerous traditions in Hinduism which affirm particular rivers, mountains, or forests as sacred. Moreover, in the concept of *līlā,* the creative play of the gods, Hindu theology engages the world as a creative manifestation of the divine. This same tension between withdrawal from the world and affirmation of it is present in Buddhism. Certain Theravāda schools of Buddhism emphasize withdrawing in meditation from the transient world of suffering (*saṃsāra*) to seek release in *nirvāṇa*. On the other hand, later Mahāyāna schools of Buddhism, such as Hua-yen, underscore the remarkable interconnection of reality in such images as the jeweled net of Indra, where each jewel reflects all the others in the universe. Likewise, the Zen gardens in East Asia express the fullness of the Buddha-nature (*tathāgatagarbha*) in the natural world. In recent years, socially engaged Buddhism has been active in protecting the environment in both Asia and the United States.

The East Asian traditions of Confucianism and Taoism remain, in certain ways, some of the most life-affirming in the spectrum of world religions.[15] The seamless interconnection between the divine, human, and natural worlds that characterizes these traditions has been described as an anthropocosmic worldview.[16] There is no emphasis on radical transcendence as there is in the Western traditions. Rather, there is a cosmology of a continuity of creation stressing the dynamic movements of nature through the seasons and the agricultural cycles. This organic cosmology is grounded in the philosophy of *ch'i* (material force), which provides a basis for appreciating the profound interconnection of matter and spirit. To be in harmony with nature and with other humans while being attentive to the movements of the *Tao* (Way) is the aim of personal cultivation in both Confucianism and Taoism. It should be noted, however, that this positive worldview has not prevented environmental degradation (such as deforestation) in parts of East Asia in both the premodern and modern period.

In a similar vein, indigenous peoples, while having ecological cosmologies have, in some instances, caused damage to local environments through such practices as slash-and-burn agriculture. Nonetheless, most indigenous peoples have environmental ethics embedded in their worldviews. This is evident in the complex reciprocal obligations surrounding life-taking and resource-gathering which mark a community's relations with the local bioregion. The religious views at the basis of indigenous lifeways involve respect for the sources of food, clothing, and shelter that nature provides. Gratitude to the creator and to the spiritual forces in creation is at the heart of most indigenous traditions. The ritual calendars of many indigenous peoples are carefully coordinated with seasonal events such as the sound of returning birds, the blooming of certain plants, the movements of the sun, and the changes of the moon.

The difficulty at present is that for the most part we have developed in the world's religions certain ethical prohibitions regarding homicide and restraints concerning genocide and suicide, but none for biocide or geocide. We are clearly in need of exploring such comprehensive cosmological perspectives and communitarian environmental ethics as the most compelling context for motivating change regarding the destruction of the natural world.

Responses of Religions to the Environmental Crisis

How to chart possible paths toward mutually enhancing human-earth relations remains, thus, one of the greatest challenges to the world's religions. It is with some encouragement, however, that we note the growing calls for the world's religions to participate in these efforts toward a more sustainable planetary future. There have been various appeals from environmental groups and from scientists and parliamentarians for religious leaders to respond to the environmental crisis. For example, in 1990 the Joint Appeal in Religion and Science was released highlighting the urgency of collaboration around the issue of the destruction of the environment. In 1992 the Union of Concerned Scientists issued the statement "Warning to Humanity," signed by over 1,000 scientists from 70 countries, including 105 Nobel laureates, regarding the gravity of the environmental crisis. They specifically cited the need for a new ethic toward the earth.

Numerous national and international conferences have also been held on this subject and collaborative efforts have been established. Environmental groups such as World Wildlife Fund have sponsored interreligious meetings such as the one in Assisi in 1986. The Center for Respect of Life and Environment of the Humane Society of the United States has also held a series of conferences in Assisi on

Spirituality and Sustainability and has helped to organize one at the World Bank. The United Nations Environmental Programme in North America has established an Environmental Sabbath, each year distributing thousands of packets of materials for use in congregations throughout North America. Similarly, the National Religious Partnership on the Environment at the Cathedral of St. John the Divine in New York City has promoted dialogue, distributed materials, and created a remarkable alliance of the various Jewish and Christian denominations in the United States around the issue of the environment. The Parliament of World Religions held in 1993 in Chicago and attended by some 8,000 people from all over the globe issued a statement of Global Ethics of Cooperation of Religions on Human and Environmental Issues. International meetings on the environment have been organized. One example of these, the Global Forum of Spiritual and Parliamentary Leaders held in Oxford in 1988, Moscow in 1990, Rio in 1992, and Kyoto in 1993, included world religious leaders, such as the Dalai Lama, and diplomats and heads of state, such as Mikhail Gorbachev. Indeed, Gorbachev hosted the Moscow conference and attended the Kyoto conference to set up a Green Cross International for environmental emergencies.

Since the United Nations Conference on Environment and Development (the Earth Summit) held in Rio in 1992, there have been concerted efforts intended to lead toward the adoption of an *Earth Charter* by the year 2000. This *Earth Charter* initiative is under way with the leadership of the Earth Council and Green Cross International, with support from the government of the Netherlands. Maurice Strong, Mikhail Gorbachev, Steven Rockefeller, and other members of the Earth Charter Project have been instrumental in this process. At the March 1997 Rio+5 Conference a benchmark draft of the *Earth Charter* was issued. The time is thus propitious for further investigation of the potential contributions of particular religions toward mitigating the environmental crisis, especially by developing more comprehensive environmental ethics for the earth community.

Expanding the Dialogue of Religion and Ecology

More than two decades ago Thomas Berry anticipated such an exploration when he called for "creating a new consciousness of the multiform religious traditions of humankind" as a means toward renewal of the human spirit in addressing the urgent problems of contemporary society.[17] Tu Weiming has written of the need to go "Beyond the Enlightenment Mentality" in exploring the spiritual resources of the global community to meet the challenge of the ecological crisis.[18] While this exploration has also been the intention of both the conferences and these volumes, other significant efforts have preceded our current endeavor.[19] Our discussion here highlights only the last decade.

In 1986 Eugene Hargrove edited a volume titled *Religion and Environmental Crisis.*[20] In 1991 Charlene Spretnak explored this topic in her book *States of Grace: The Recovery of Meaning in the Post-Modern Age.*[21] Her subtitle states her constructivist project clearly: "Reclaiming the Core Teachings and Practices of the Great Wisdom Traditions for the Well-Being of the Earth Community." In 1992 Steven Rockefeller and John Elder edited a book based on a conference at Middlebury College titled *Spirit and Nature: Why the Environment Is a Religious Issue.*[22] In the same year Peter Marshall published *Nature's Web: Rethinking Our Place on Earth,*[23] drawing on the resources of the world's traditions. An edited volume titled *Worldviews and Ecology,* compiled in 1993, contains articles reflecting on views of nature from the world's religions and from contemporary philosophies, such as process thought and deep ecology.[24] In this same vein, in 1994 J. Baird Callicott published *Earth's Insights,* which examines the intellectual resources of the world's religions for a more comprehensive global environmental ethics.[25] This expands on his 1989 volumes, *Nature in Asian Traditions of Thought* and *In Defense of the Land Ethic.*[26] In 1995 David Kinsley issued a book titled *Ecology and Religion: Ecological Spirituality in a Cross-Cultural Perspective,*[27] which draws on traditional religions and contemporary movements, such as deep ecology and ecospirituality. Seyyed Hossein Nasr wrote his comprehensive study *Religion and the Order of Nature* in 1996.[28] Several volumes of religious responses to a particular topic or theme have also been published. For example, J. Ronald Engel and Joan Gibb Engel compiled a monograph in 1990 titled *Ethics of Environment and Development: Global Challenge, International Response*[29] and in 1995 Harold Coward edited the volume *Population, Consumption and the Environment: Religious and Secular Responses.*[30] Roger Gottlieb edited a useful source book, *This Sacred Earth: Religion, Nature, Environment.*[31] Single volumes on the world's religions and ecology were published by the Worldwide Fund for Nature.[32]

The series Religions of the World and Ecology is thus intended to expand the discussion already under way in certain circles and to invite further collaboration on a topic of common concern—the fate of the earth as a religious responsibility. To broaden and deepen the reflective basis for mutual collaboration was an underlying aim of the conferences themselves. While some might see this as a diversion from pressing scientific or policy issues, it was with a sense of humility and yet conviction that we entered into the arena of reflection and debate on this issue. In the field of the study of world religions, we have seen this as a timely challenge for scholars of religion to respond as engaged intellectuals with deepening creative reflection. We hope that these volumes will be simply a beginning of further study of conceptual and symbolic resources, methodological concerns, and practical directions for meeting this environmental crisis.

Notes

1. He goes on to say, "And that is qualitatively and epochally true. If religion does not speak to [this], it is an obsolete distraction." Daniel Maguire, *The Moral Core of Judaism and Christianity: Reclaiming the Revolution* (Philadelphia: Fortress Press, 1993), 13.

2. Gerald Barney, *Global 2000 Report to the President of the United States* (Washington, D.C.: Supt. of Docs. U.S. Government Printing Office, 1980–1981), 40.

3. Lynn White, Jr., "The Historical Roots of Our Ecologic Crisis," *Science* 155 (March 1967):1204.

4. Thomas Berry, *The Dream of the Earth* (San Francisco: Sierra Club Books, 1988).

5. Brian Swimme and Thomas Berry, *The Universe Story* (San Francisco: Harper San Francisco, 1992).

6. At the same time we recognize the limits to such a project, especially because ideas and action, theory and practice do not always occur in conjunction.

7. E. N. Anderson, Ecologies of the Heart: Emotion, Belief, and the Environment (New York and Oxford: Oxford University Press, 1996), 166. He qualifies this statement by saying, "The key point is not religion per se, but the use of emotionally powerful symbols to sell particular moral codes and management systems" (166). He notes, however, in various case studies how ecological wisdom is embedded in myths, symbols, and cosmologies of traditional societies.

8. *Is It Too Late?* is also the title of a book by John Cobb, first published in 1972 by Bruce and reissued in 1995 by Environmental Ethics Books.

9. Because we cannot identify here all of the methodological issues that need to be addressed, we invite further discussion by other engaged scholars.

10. See J. Baird Callicott, *Earth's Insights: A Survey of Ecological Ethics from the Mediterranean Basin to the Australian Outback* (Berkeley: University of California Press, 1994).

11. See, for example, The Quality of Life, ed. Martha C. Nussbaum and Amartya Sen, WIDER Studies in Development Economics (Oxford: Oxford University Press, 1993).

12. White, "The Historical Roots of Our Ecologic Crisis," 1203–7.

13. Process theology, creation-centered spirituality, and ecotheology have done much to promote these kinds of holistic perspectives within Christianity.

14. These are resources already being explored by theologians and biblical scholars.

15. While this is true theoretically, it should be noted that, like all ideologies, these traditions have at times been used for purposes of political power and social control. Moreover, they have not been able to prevent certain kinds of environmental destruction, such as deforestation in China.

16. The term "anthropocosmic" has been used by Tu Weiming in *Centrality and Commonality* (Albany: State University of New York Press, 1989).

17. Thomas Berry, "Religious Studies and the Global Human Community," unpublished manuscript.

18. Tu Weiming, "Beyond the Enlightenment Mentality," in *Worldviews and Ecology,* ed. Mary Evelyn Tucker and John Grim (Lewisburg, Pa.: Bucknell University Press, 1993; reissued, Maryknoll, N.Y.: Orbis Books, 1994).

19. This history has been described more fully by Roderick Nash in his chapter entitled "The Greening of Religion," in The Rights of Nature: A History of Environmental Ethics (Madison: University of Wisconsin Press, 1989).

20. *Religion and Environmental Crisis,* ed. Eugene Hargrove (Athens: University of Georgia Press, 1986).

21. Charlene Spretnak, *States of Grace: The Recovery of Meaning in the Post-Modern Age* (San Francisco: Harper San Francisco, 1991).

22. *Spirit and Nature: Why the Environment Is a Religious Issue,* ed. Steven Rockefeller and John Elder (Boston: Beacon Press, 1992).

23. Peter Marshall, *Nature's Web: Rethinking Our Place on Earth* (Armonk, N.Y.: M. E. Sharpe, 1992).

24. *Worldviews and Ecology,* ed. Mary Evelyn Tucker and John Grim (Lewisburg, Pa.: Bucknell University Press, 1993; reissued, Maryknoll, N.Y.: Orbis Books, 1994).

25. Callicott, *Earth's Insights.*

26. Both are State University of New York Press publications.

27. David Kinsley, *Ecology and Religion: Ecological Spirituality in a Cross-Cultural Perspective* (Englewood Cliffs, N.J.: Prentice Hall, 1995).

28. Seyyed Hossein Nasr, *Religion and the Order of Nature* (Oxford: Oxford University Press, 1996).

29. *Ethics of Environment and Development: Global Challenge, International Response,* ed. J. Ronald Engel and Joan Gibb Engel (Tucson: University of Arizona Press, 1990).

30. *Population, Consumption, and the Environment: Religious and Secular Responses,* ed. Harold Coward (Albany: State University of New York Press, 1995).

31. This Sacred Earth: Religion, Nature, Environment, ed. Roger S. Gottlieb (New York and London: Routledge, 1996).

32. These include volumes on Hinduism, Buddhism, Judaism, Christianity, and Islam.

Introduction

CHRISTOPHER KEY CHAPPLE

The Jain Faith in History

The Jain religion originated more than twenty-five hundred years ago in India. It developed a path of renunciation and purification designed to liberate one from the shackles of *karma*, allowing one to enter into a state of eternal liberation from rebirth, or *kevala*, which is roughly equivalent to the Buddhist concept of *nirvāṇa*. The primary method of attaining this ultimate state requires a careful observance of nonviolent behavior. Jainism emphasizes nonviolence, or *ahiṃsā*, as the only true path that leads to liberation and prescribes following scrupulous rules for the protection of life in all forms.[1]

The origins of Jainism are somewhat difficult to trace. The tradition holds that twenty-four great teachers, or Tīrthaṅkaras, established the foundations of the Jain faith. The most recent of these teachers, Vardhamāna Mahāvīra (also known as the Jina) most probably lived during the time of the Buddha. Recent scholarship suggests that the Buddha lived in the fourth century B.C.E. However, the traditional stories of Mahāvīra indicate that he was born into a family that followed the religious teachings of Pārśvanātha, the twenty-third Tīrthaṅkara, who possibly taught during the eighth century B.C.E. Because virtually no archaeological ruins can be found in India for the period from 1500 to 300 B.C.E., exact dates cannot be determined. However, the first excavations of northern India during the Hellenistic era (ca. 300 B.C.E.) include statues of Jain images. Furthermore, the earliest Buddhist texts discuss Jainism in some detail, suggesting that it was a well-established tradition even before the time of the Buddha.

The records of Strabo (64 B.C.E. to 23 C.E.), the Greek geographer, describe two prevailing styles of religiosity in India at the time of Alexander (ca. 330 B.C.E.), as recorded by Megasthenes (350–290 B.C.E.): the Brahmanical traditions, later described by the Persians as "Hindu," and the Śramaṇical traditions, which include Buddhism and Jainism.[2] The Brahmanical traditions emphasize the Vedas, ritual, and the authority of a priestly caste. The Śramaṇical traditions do not accept the Vedas, advocate meditation rather than ritual, and look to monks and nuns for religious authority. Buddhism sent out missionaries from India who established Theravāda Buddhism in Southeast Asia, Mahāyāna Buddhism in East Asia, and Vajrayāna Buddhism in Central Asia. Buddhism flourished in India until the tenth century, when its influence waned.

Jainism did not establish a missionary tradition but cultivated a strong laity. Like Buddhism, it began in Northeast India but, possibly because of drought in the third century B.C.E., many Jains moved to the southern kingdoms of Karnataka and Tamil Nadu, as well to the western parts of India now known as Gujarat, Rajasthan, and Madhya Pradesh. Eventually, two sects of Jainism arose: the Digambaras, primarily found in central and southern India, and the Śvetāmbaras, who live primarily in western India. The two groups agree on the foundational Jain principles of *karma* and nonviolence. However, they differ on their biographical accounts of Mahāvīra, accept different texts as authentically canonical, and hold divergent views on renouncing clothing and on the potential spiritual status of women. The Śvetāmbaras, whose name means "white clad," contend that monks and nuns can achieve the highest levels of spirituality without renouncing their clothing. They also believe that women hold the potential to achieve the state of liberation, or *kevala*. The Digambaras, whose name means "sky clad," hold that all clothing must ultimately be renounced and that, because only men are allowed to take this ultimate vow of renunciation, a woman must be reborn as a man to achieve *kevala*. These traditions arose in geographic isolation from one another and developed into distinct schools by the early centuries of the common era.

The *Ācārāṅga Sūtra* (ca. 400 B.C.E.), a text used extensively by the Śvetāmbaras, is the oldest surviving Jain manual, describing the rules proclaimed by Mahāvīra to be followed by his monks and nuns. One thinker, Umāsvāti, who probably lived in the fourth century C.E., developed a philosophical approach to Jainism that both Śvetāmbaras and Digambaras accept. In a text known as the *Tattvārtha Sūtra*, or *Aphorisms on the Meaning of Reality*, he succinctly outlines the Jain worldview, describing *karma*, cosmology, ethics, and the levels of spiritual attainment (*guṇasthāna*). Later philosophers, including Haribhadra (ca. 750 C.E.) and Hemacandra (ca. 1150 C.E.) of the Śvetāmbara tradition and Jinasena (ca. 820 C.E.) and Vīrasena (ca. 800 C.E.) of the Digambara tradition, developed an extensive literary corpus that includes stories, epics, philosophical treatises, and poetry. During the Mogal period, Jinacandrasūri II (1541–1613), the leader of the Kharatara Gaccha (a subdivision of the Śvetāmbara sect) achieved great influence at the court of Akbar, convincing the emperor to protect Jain pilgrimage places. Akbar even prohibited animal slaughter for one week per year under Jinacandrasūri's urging. In contemporary times, Jain have become very influential in the areas of publishing, law, and business. They continue to work at integrating their philosophy of nonviolence into the daily life of India.

The Jain community has also participated in an extensive diaspora, with several tens of thousands living in various parts of the world. Jain business families settled in East Africa several decades ago. After Indian independence, some Jains settled in Great Britain, with a great influx from East Africa during the expulsion of all South Asians from Uganda under the rule of Idi Amin. In Kobe, Japan, Jains participate in the diamond trade. Jains began migrating to North America after the changes in immigration law in 1965, inspired by the Civil Rights movement. These new immigrants have built temples and organized several networks and organizations for maintaining Jain identity, including the Jaina Associations in North America (JAINA), which sponsors semiyearly conventions. These gatherings have included presentations pertaining to current issues, such as environmentalism.[3]

Jainism and Environmentalism

The common concerns between Jainism and environmentalism can be found in a mutual sensitivity toward living things, a recognition of the interconnectedness of life-forms, and support of programs that educate others to respect and protect living systems. For the Jains, this approach is anchored in a cosmology that views the world in terms of a cosmic woman whose body contains countless life souls (*jīva*) that reincarnate repeatedly until the rare attainment of spiritual liberation (*kevala*). The primary means to attain freedom requires the active nonharming of living beings, which disperses the *karma*s that keep one bound. Jains adhere to the vows of nonviolence to purify their *karma* and advance toward the higher states of spiritual attainment (*guṇasthāna*). For Jain laypeople, this generally means keeping to a vegetarian diet and pursuing livelihoods deemed to inflict a minimum of harm. For Jain monks and nuns, this means the need to avoid doing harm to all forms of life, including bugs and microorganisms (*nigoda*).

Contemporary environmental thinkers in the developed world, particularly within the last decade of the twentieth century, have come to emphasize the interconnectedness of life as the foundation for developing an environmental ethic. On the policy level, the Endangered Species Act of the United States extends protection to even the smallest aspect of life, emphasizing the microphase as the key to ecosystem protection. Taking a different approach, Norway has developed a comprehensive approach to assess the impact of one action on the broader network of relationships within a given biome.[4] Both approaches grapple with the age-old problem of how to balance the needs of the one and the many when working toward the highest good.

Drawing from her own relationships with trees, ecologist Stephanie Kaza has proposed an approach to the natural world that engenders feelings of tenderness, respect, and protection. She writes:

> The relationship between person and tree, arising over and over again in many different contexts and with various individuals, is one subset of all human-nonhuman relationships. . . . I want to know, What does it actually mean to be in a relationship with a tree? Acknowledgment of and participation in relationships with trees, coyotes, mountains, and rivers is central to the philosophy of deep ecology. . . . In the course of studying mountains and rivers in depth, one sees them explode into all the phenomena that support their existence—clouds, stones, people walking, animals crawling, the earth shaking.[5]

By participating in the close observation of individual life processes, in this case using the tree as a starting point, one begins to see the network of relationships that enlivens all forms of consciousness. By gaining intimacy with a small part of the whole, concern for the larger ecosystem arises. Each piece, no matter how small, contributes to the whole. To disrupt the chain of life at any link can result in dire consequences, as seen in the release of radioactivity in Chernobyl, the great industrial accident in Bhopal, the depletion of the ozone layer over the polar caps, and the extinction of various species of plants and animals.

As seen in the above example from Stephanie Kaza, an important impetus for environmental activism comes from the close observance and consequent appreciation of the external world. As our ecosystem becomes impoverished, humans take notice and respond. Ultimately, this concern for nature can be seen as a form of self-preservation, as the earth is the only context for human flourishing. Similarly, according to the *Ācārāṅga Sūtra*, Mahāvīra was moved when he observed nature at close range, noticing that even the simplest piece of a meadow teems with life:

> Thoroughly knowing the earth-bodies and water-bodies and fire-bodies and wind-bodies, the lichens, seeds, and sprouts, he comprehended that they are, if narrowly inspected, imbued with life. . . .[6]

In a contemporary echo of this realization, James Laidlaw records the conversion moment of a woman who subsequently decided to become a Jain nun:

> the decision came one morning when she walked into the kitchen. There was a cockroach in the middle of the floor, "and I just looked at it and suddenly I thought, 'Why should I stay in this world where there is just suffering and death and rebirth?'"[7]

Seeing the life and spirit of a lowly insect inspired this woman to pursue a lifelong commitment of harmlessness to all beings. Benevolence to souls other than one's own leads to self-purification and the transcendence of worldly entanglements. The ethics of nonviolence as developed by the Jains looks simultaneously inward and outward. The only path for saving one's own soul requires the protection of all other possible souls.

Jainism offers a worldview that in many ways seems readily compatible with core values associated with environmental activism. While both uphold the protection of life, the underlying motives governing the Jain faith and those governing environmental activism do differ. First, as various authors in this book will point out, the telos or goal of Jainism lies beyond all worldly concerns. The Jain observances of nonviolence, for instance, are not ultimately performed for the sake of protecting the individual uniqueness of any given lifeform for its own sake. The reason for the protection of life is for self-benefit, stemming from a desire to avoid accruing a karmic debt that will result in later retribution against oneself. The result may be the same; a life might be spared. However, this is a by-product of a desire to protect and purify oneself through the avoidance of doing harm. In the case of some environmental activists, aggressive, direct action might be undertaken to interfere with and stop the destruction of a natural habitat in a way that might be seen as violent, such as the monkey-wrenching techniques used by EarthFirst![8] This would not be acceptable to a Jain.

In this volume the following questions will be posed: How does traditional Jain cosmology, and its consequent ethics, view the natural world? Is this worldview compatible with contemporary ecological theory? How might a Jain ethical system respond to the challenges of making decisions regarding such issues as the development of dams, the proliferation of automobiles, overcrowding due to overpopulation, and the protection of individual animal species? Can there be a Jain environmental activism that stems from a traditional concern for self-purification that simultaneously responds to the contemporary dilemma of ecosystem degradation?

In the chapters that follow, this topic will be pursued from a variety of perspectives. The voices included in this volume reflect a wide spectrum of approaches. Several scholars born and trained in the West take a critical look at the real prospects for Jain advocacy of environmental protection. Jain scholars from India, on the other hand, see actual solutions in Jain philosophy for correcting ecological imbalances through a reconsideration of lifestyle and active application of *ahiṃsā*. Perhaps the closest analogue to environmental activism within historical Jainism can be found in the tradition of animal protection, as found in the many hundreds, if not thousands, of shelters, or *pinjrapoles*, located in and near Jain communities in western India.[9] Modern initiatives, some of which are mentioned in this book, include tree-planting prgrams at pilgrimage sites. Dr. Michael Fox of the Humane Society and the Center for Respect of Life and Environment has re-energized an animal shelter inspired by Jain values in South India.[10] By combining the ancient practice of animal protection with considered reflections on how traditional Jain observances of nonviolence might counter the excesses of the modern, industrialized, consumer-oriented lifestyle, the Jain faith might provide a new voice for the development of ecofriendly behaviors.

Overview of the Volume

The book has been divided into four sections, followed by an appendix and a bibliography. The first section examines Jain theories about the nature of the universe, which then provide the context for developing an ecological interpretation of the tradition. The second section raises some challenges to the possibility of developing an ecofriendly Jain ethic. The third section, written by Jain practitioners, asserts that Jainism, with its emphasis on nonviolence (*ahiṃsā*), is inherently sensitive to and practically responsive to environmental needs. The fourth section discusses the adaptation of ecological ideas among select members of the contemporary Jain community, largely among its diaspora adherents.

In the first chapter, Nathmal Tatia, who passed away shortly after the conference on Jainism and ecology took place in the summer of 1998, suggests that virtually all the religious traditions of the world "contain aspects that are not anthropocentric" and then introduces key aspects of Jain philosophy. Noting that neither Jainism nor Buddhism contains a creating or controlling God, he emphasizes compassion as the key for the protection of life. Tatia suggests that the Jain advocacy of vegetarianism and protection of animals provide a possible remedy for the current ecological crisis. He provides a synoptic view of how the application of traditional Jain ethics can help one enact environmentalist values.

Philosopher John Koller probes the Jain theory of many-sidedness (*anekānta*) as an antidote to the one-theory approach that drives the development machine and has led to environmental degradation. Jains traditionally seek to understand any situation from as many angles as possible, as exemplified in the famous story of the six blind men and the elephant. One feels the tail and "sees" a snake. Another feels the ear and "sees" a fan, and so forth. Each can claim a "truth," but no one, at least before the experience of *kevala*, can claim to see totality. By utilizing a multiple-perspective approach to environmental issues, Koller suggests that Jains will be better equipped to cope with such ethical dilemmas as the use and abuse of trees and oceans.

Kristi Wiley begins her chapter with an assessment of the discipline of environmental ethics as it has evolved in Western academia. Noting the shift from anthropocentrism to biocentrism, Wiley sees some commonalities between the moral considerations of Jainism and systems ecologists. Her careful interpretation of indigenous Jain biology and elemental theory lists in detail the karmic effects of negative interactions with one's environment. She makes the important distinction between beings with consciousness (*saṃjñī*) and those without consciousness (*asaṃjñī*), which provides some basis for using plants and the elements as resources for human sustenance. Wiley also emphasizes the central role played by the nuns and monks who serve as the conscience of the Jain tradition, advocating protection for even those beings who lack awareness, such as plants and the living bodies contained within earth, water, fire, and air.

The second section poses challenges to the conventional assumption that Jainism by its very nature contains all the precepts of environmentalism. It begins with an essay by John Cort, who suggests that a great deal of work needs to be accomplished before the Jain tradition can honestly claim to be ecofriendly. Noting that the environmental crisis is a recent development, he suggests that environmental thought and activism might help inform how Jains define and realize their commitment to *ahiṃsā*. In particular, he discusses the Jain "value of wellbeing" as providing a counterbalance to the Jain emphasis on liberation, noting that "Jain ethics . . . are highly context-sensitive" and hence adaptable according to time and place. He compares and contrasts ecofeminism and the role of women in Jainism, and suggests that social ecology must be taken into consideration, noting that the project to reforest Jain pilgrimage sites has had a negative effect on low-caste herders whose livestock have become restricted from foraging. Acknowledging the long history of Jainism as a social catalyst, Cort looks forward to the development of a "distinctive Jain environmental ethic."

Paul Dundas suggests that in the history of Jainism some attitudes toward nature may have been less than ecofriendly. He describes the dualistic and pluralistic nature of Jain philosophy, which divides the world into living and nonliving entities, with each living entity (*jīva*) responsible for its own fate. Dundas states that within this worldview nature in and of itself has no "autonomous value." Value lies in the human application of nonviolence to attain, as noted earlier in this introduction, the release of all *karma* and the eventual severance from all materiality, including "nature." To apply purely monastic values to the issue of ecological degradation simply does not work, argues Dundas, citing various ethical tales about elephant-eating ascetics, brutal horse tamers, and well diggers, each of which seems to contain, at best, an ambiguous environmental ethic. He cautions that one must exert care in attempting to match a "traditional soteriological path" to "fit the requirements of a modern, ultimately secular, Western-driven agenda."

My own chapter suggests that the Jain community could benefit from examining its worldview and ethics in light of some contemporary theorists in the area of religion and ecology, specifically Brian Swimme, Thomas Berry, and David Abram. Each of these three has highlighted the dynamic aspects of living processes, displaying a sensitivity to life somewhat similar to that found in Jainism. David Abram has emphasized in particular the role of the senses in determining and defining reality, taking an approach comparable to the empiricism emphasized in Umāsvāti's *Tattvārtha Sūtra*, the Buddhist Abhidharma schools, and the Hindu schools of Sāṃkhya and Yoga. The Jain worldview that sees the universe, from earth-bodies to human beings, as suffused with life accords with the thought of Thomas Berry, who has stated that the world is a "communion of subjects, not a collection of objects." Furthermore, the Jain assertion that even the earth itself feels our presence is strikingly resonant with the observations of Brian Swimme. The pan-psychic vision of Jainism is compared and contrasted with contemporary Western scientific and philosophical insights, with the suggestion that these two fields be brought into closer dialogue with one another.

Padmanabh S. Jaini, one of the world's leading scholars of Jainism, summarizes fundamental Jain teachings and then seeks to explore how Jainism might respond to key issues of development and economics. The current drive toward industrialization and consumerism in India violates many essential Jain precepts, particularly nonpossession (*aparigraha*). By examining traditional lifestyles and occupations, as well as Jain attitudes toward wealth in general, Jaini suggests that a balanced approach to development can be pursued.

In the third section of the book, Jain practitioners suggest that Jainism already has developed a working environmental ethics. As such, this section represents an emic, or insider's, view of Jainism. It includes three essays that might fit more within the genre of a sermon than an academic paper, but which nonetheless make an important contribution to this emerging discourse. These chapters point to new directions to be taken within the practice of Jainism, grounded in the earlier tradition.

Sadhvi Shilapi, a prominent Jain nun, raises up the voice of Mahāvīra, the great Jain Tīrthaṅkara of twenty-five hundred years ago, to suggest how Jains can and should respond to the problems of industrialization, population growth, and human exploitation of nonhuman life-forms. Quoting from the *Ācārāṅga Sūtra*, the oldest text of the Śvetāmbara Jain tradition, she suggests that Mahāvīra's sensitivity to plants and the elements themselves can serve to inform the Jain response to resource limitations. She also emphasizes the need for tree planting in rural areas of India, an initiative taken by her own religious community, Veerayatan, in Bihar.

Bhagchandra Jain consults a wide range of Jain literature from both the Śvetāmbara and Digambara schools to compile a masterful argument for the respect of all life-forms. He notes the extensive literature within Jainism devoted to forest protection and emphasizes the ecological aspects of behavior recommended for Jain laypersons.

Satish Kumar, founder and educational director of Schumacher College in England, relates the concept of ecology to the simple lifestyle observed by his own mother, which included strict vegetarianism, pilgrimages to sacred mountains, constant observance of barefootedness, minimalization of possessions, conservation of water, and close adherence to an ethical code grounded in nonviolence.

In the concluding section, Anne Vallely examines the tensions between traditional and contemporary Jainism, particularly in its current globalized form. She notes the trend by some Jains to identify themselves as ecofriendly. She then examines what she terms to be a newly emerging sociocentric ecological worldview within the Jain community. Diaspora Jains, particularly in North America, have brought about a distinctive form of Jainism that emphasizes "the values of vegetarianism, animal welfare, meditation, and active promotion of interfaith activities." Though the inspiration of each of these can be seen as having its roots in Jain thought and practice, they are being played out in a far more public arena than that traditionally observed by the inward Jain ascetics.

The volume concludes with an appendix, *The Jain Declaration on*

Nature, prepared by L. M. Singhvi, a member of the Indian Parliament and former high commissioner from India to the United Kingdom. This was originally published as a small booklet in 1992. This document has helped stimulate the discussion of environmental values in the Jain community worldwide and serves as an example of what Anne Vallely refers to as the newly emerging sociocentric expression of Jainism.

Contemporary Theorists of Jain Ecology

The interface between Jainism and ecology remains a complex issue, and it is important to recognize some of the pioneers in this emerging discussion. Though he was not able to participate in the Harvard conference, the work and commitment of Michael Tobias must be acknowledged. Tobias, who received his doctorate in the history of consciousness, has worked for several decades as a writer and filmmaker dedicated to environmental causes. In 1988 he released the film "Ahimsa," which elegantly portrays several Jain leaders and extols the religion as the great champion of animal rights and nonviolent living. He wrote a book titled *Life Force: The World of Jainism* that serves as a written companion to the film, and he contributed the chapter on Jainism to Mary Evelyn Tucker and John Grim's *Worldviews and Ecology*.[11] Though not trained as a scholar of Jainism, Tobias nonetheless recognized a commonality between his own environmental interests and the Jain worldview. He remains a sought-after speaker within the extensive network of Jain conferences and proclaims himself to be a Jain.

The work of Satish Kumar, both with his journal *Resurgence* and the curriculum that he has developed at Schumacher College, indicates his willingness to blend together social activism and a Jain-inspired commitment to nonviolence. Kumar left the life of a traditional Jain monk to join the land redistribution movement of Vinobha Bhave (1895–1982), and later journeyed as a peace activist on foot from Delhi to Moscow to Paris in an attempt to stop nuclear proliferation in the 1960s.[12] He has most recently joined forces with Dr. Atul K. Shah to produce the journal *Jain Spirit: Advancing Jainism into the Future*, which is published six times each year and distributed internationally. Each issue includes articles and photo essays that reinforce an ecofriendly view. Most of the articles in the "Environment" section of the magazine are by environmental activists such as David Ehrenfield, Joyce D'Silva, and Donella Meadows and serve more to educate Jains about contemporary trends in the field of ecology than to articulate a distinctly Jain vision of environmentalism. Kumar has attempted a synthesis of spirituality and activism, inspired in part by his childhood and young adult years as a monk in Ācārya Tulsi's Śvetāmabara Terāpanthī movement, which includes ten special vows that were formulated in 1949, including "I will always be alert to keeping the environment pollution-free."

The Advent of Jain Environmentalism

As Anne Vallely notes in her chapter, some modern Jains, particularly in North America, see involvement with environmental causes and animal rights activism as a logical extension of their faith. However, how authentic is this tradition? Is it, as Vallely suggests, a revision of asceticism? Can the observance and advocacy of vegetarianism and ecological sensitivity substitute as a new form of asceticism? Can Jainism truly survive without the living presence of monks and nuns to chide and inspire the more worldly lay community?

In the modern diaspora context, traditional monasticism, rigorously practiced by monks and nuns in India, has not taken root, nor does it seem to be a likely option, given the relatively small numbers of Jains living outside India and the logistical difficulties of providing the donor support sanctioned by the Jain lay community. However, some Jain monastics (and former Jain monastics), such as Muni Sri Chitrabhanu, Acharya Sushil Kumar, Sadhvi Shilapi, and Satish Kumar, have helped promulgate Jain teachings outside of India, and many nuns in training (*samani*s) from the Terāpanthī community have lectured throughout the world. Numerous lay Jains participate in regular practices of fasting and other austerities, particularly the Paryūṣaṇa fast observed in late August. The many Jain centers and temples throughout North America and the United Kingdom have developed extensive weekend educational programs for children (*pāṭśāḷās*), camps, retreats, and web sites to educate their members (and others) about the faith. Many of these activities include mention of the environment from a Jain perspective.

This volume points to the dynamic nature of the Jain faith and its willingness to engage in discussion on this modern social issue. Not unlike nearly any other religious tradition, it remains to be seen if the Jain worldview and ethic can inspire an effective ecological vision. Can Jainism adopt a sociocentric environmental point of view without compromising its core values? Hopefully, this collection of essays will help advance this discussion.

Notes

1. For information on the history, philosophy, and practice of Jainism, see Padmanabh S. Jaini, *The Jaina Path of Purification* (Delhi: Motilal Banarsidass, 1979); Paul Dundas, *The Jains* (London: Routledge, 1992); Alan Babb, *Absent Lord: Ascetics and Kings in Jain Ritual Culture* (Berkeley: University of California Press, 1996); and John E. Cort, *Jains in the World: Religious Values and Ideology in India* (New York: Oxford University Press, 2001).

2. *The Geography of Strabo*, trans. Horace Leonard Jones (New York: Putnam, 1930), 101.

3. See Marcus Banks, *Organizing Jainism in India and England* (Oxford: Clarendon Press, 1992). See also *Peace through Non-violence: Eighth Biennial Jaina Convention Souvenir Volume* (Chicago: Federation of Jain Associations in North America, 1995); and Bhuvanendra Kumar, *Jainism in America* (Mississauga, Ontario: Jain Humanities Press, 1996).

4. See David Rothenberg, "Individual or Community? Two Approaches to Ecophilosophy in Practice," in *Ecological Prospects: Scientific, Religious, and Aesthetic Perspectives*, ed. Christopher Key Chapple (Albany: State University of New York, 1994), 83–92.

5. Stephanie Kaza, *The Attentive Heart: Conversations with Trees* (New York: Fawcett Columbine, 1993), 10–11.

6. *Ācārāṅga Sūtra* 1.8.1.11–12; from *Jaina Sutras*, Part 1, *The Ākārāṅga Sūtra. The Kalpa Sūtra*, trans. Hermann Jacobi (1884; New York: Dover, 1968).

7. James Laidlaw, *Riches and Renunciation: Religion, Economy, and Society among the Jains* (Oxford: Clarendon Press, 1995), 157.

8. See "Earth First! And Global Narratives of Popular Ecological Resistance," in *Ecological Resistance Movements: The Global Emergence of Radical and Popular Environmentalism*, ed. Bron Raymond Taylor (Albany: State University of New York Press, 1995), 11–34.

9. Deryck O. Lodrick, *Sacred Cows, Sacred Places: Origins and Survivals of Animal Homes in India* (Berkeley: University of California Press, 1981).

10. See the web site for the India Project for Animals and Nature: www.gcci.org.

11. *Worldviews and Ecology*, ed. Mary Evelyn Tucker and John Grim (Maryknoll, N.Y.: Orbis Books, 1994).

12. See Satish Kumar's autobiography, *Path without Destination* (New York: William Morrow, 1999).

Jain Theories about the Nature of the Universe

The Jain Worldview and Ecology

NATHMAL TATIA

The Jain Worldview

Jainism has a system of ethics that places nonviolence (*ahiṃsā*) at the top of its principles of morality. Abstaining from any kind of injury, in thought, word, and deed, to any kind of living being, immobile or mobile, is *ahiṃsā*.[1] As is explained in other chapters in this volume, Jainism lists six classes of living beings, namely, earth-bodied, water-bodied, air-bodied, fire-bodied, vegetation which has only the sense of touch, and mobile beings that are two-sensed, three-sensed, four-sensed, and five-sensed, with or without the mind, which is the instrument of thought.

Humanity is the most developed class of living beings, but, according to Jainism, is not a privileged species in any sense. Human reality is not the center of things, with the right to conquer and subdue nature, just as in modern astronomy the earth is not the center of the universe with the planets, the stars, and the sun all circling round it. Even outside of India, human self-importance has been diminished by Darwin, who showed that human beings have slowly evolved and are an integral part of creation. Freda Rajotte writes:

> It is beginning to look as though humanity is just a small part of a much larger living body referred to as the 'biomass' or 'biosphere'—the whole earth and the life on it.
>
> Just as in a human body you cannot separate the life of the lungs from the life of the stomach or the life of the brain, it is beginning to look as though you cannot separate the life of human beings from the life of the trees, the insects or the seas."[2]

In Jainism raw soil, raw minerals, snow, ice, rain, vegetation, and so on are considered living organisms that are a class of immobile beings,[3] though we do not find any concept of the entire biosphere as a living body.[4] Jainism does not profess a unified or monistic worldview but emphasizes the individual integrity of each living soul.

Jainism, like Hinduism and Buddhism, is centered on life, not on the human person. These traditions differ from the religions that regard humankind to have been created by God in his divine image. Jainism gives reverence to all life, irrespective of its place in the biological hierarchy.

Ecology in Judaism, Christianity, and Islam

In Judaism, says Norman Solomon, "humankind is superior to animals, animals to plants, plants to the inanimate. There is a hierarchy in created things." He continues:

> The hierarchical model has two practical consequences. First . . . is that of responsibility of the higher for the lower, traditionally expressed as 'rule', latterly as 'stewardship'. The second is that, in a competitive situation, the higher has priority over the lower. Humans have priority over dogs so that, for instance, it is wrong for a man to risk his life to save that of a dog though right, in many circumstances, for him to risk his life to save that of another human."[5]

Thus, Judaism "values human life more than that of other living things, but at the same time stresses the special responsibility of human beings to 'work and look after' the created order."[6]

Christianity arose from the joining of Hellenistic thought and the Jewish tradition as newly interpreted by Jesus. Jesus' teachings did not directly address issues of nature or the treatment of animals. However, his teachings on love might be expanded to include more than just humans. Jesus proclaimed his message of love in the Sermon on the Mount: "Do not set yourself against the man who wrongs you. If someone slaps you on the right cheek, turn and offer him your left. . . . Love your enemies and pray for your persecutors; only so can you be children of your heavenly father, who makes his sun rise on good and bad alike, and sends the rain on the honest and the dishonest. If you love only those who love you, what reward can you expect?"[7] Jesus used nature themes frequently in his parables to remind his disciples of God's beneficence. In what might be considered a rebuke against consumerism, he warned his followers to "put away anxious thoughts about food and drink to keep you alive, and clothes to cover your body. Surely life is more than food, the body more than clothes."[8] He commented on the birds who thrive without hoarding food and the splendid beauty of the lilies who neither work nor spin. Jesus often referred to the world of nature in his teachings.

The primary text of Islam, the Qur'ān, puts the human species in the same class as the other species, by describing humans as two-legged animals: "God has created every animal out of water: of them there are some that creep on their bellies; others that walk on two legs; and others on four."[9] According to Islam, each individual and each species is a part of life as a whole. The Holy Prophet advises us to "Live in this world as if you are going to live for ever: prepare for the next world as if you were going to die tomorrow."[10] The Holy Prophet related that once upon a time, a prophet "was stung by an ant and ordered the whole of the ants' nest to be burnt. At this, God reprimanded him in these words: 'Because one ant stung you, you burned a whole community that glorified Me.'"[11] Clearly, Muhammed saw even the lowly insects as part of God's creation and hence worthy of respect.

All these traditions contain aspects that are not anthropocentric. God allots to humankind the duty to render service to all creatures. God commands his people through the prophets to respect nature, to learn from nature, and to protect nature.

Protection of Nature in Nontheistic Traditions: Friendliness and Compassion

In Jainism, which does not posit a controlling creator God, each soul is an autonomous unit struggling for salvation in its own way. Similarly, in Buddhism, each *bodhisattva* is devoted to the spiritual well-being of others and refuses to enter *nirvāṇa* leaving any single living being behind him entangled in misery and suffering. The Buddhists and Jains do not depend upon God, but the whole force of their religion is directed to the central truth in every religion that each person needs to remake himself or herself in the image of the divine ideal.[12]

The compassion (*anukampā*) of Vardhamāna Mahāvīra is a many-faceted virtue. It is abstinence from causing suffering to any living thing, on the one hand, and the positive act of rendering service to others for eradicating their suffering, on the other hand. This is the implication of *Tattvārtha Sūtra* 7.1, which prescribes abstinence from violence; 6.13, which prescribes compassion through charity; and 5.21, which defines the function of souls as rendering service to one another.[13]

For the Buddhists, friendliness (*maitri*) and compassion (*karuṇa*) as exemplified by Gautama Buddha are the two most important qualities for a spiritual aspirant to develop in life. Friendliness

> is understood as the wish for others to be happy, and compassion as the wish to alleviate suffering. Both start with ourselves, by recognizing the fact of our own suffering and seeking to uproot its causes. Before turning to the plight of others, it is necessary to understand deeply the origins of suffering within ourselves. Such insight can then lead to a genuine capacity to show others the way to freedom from their inner pain. Ultimately [friendliness] and compassion extend to all living things: people, animals, plants, the earth itself.[14]

Ahiṃsā as the Central Theme of the Jain Canon

According to Jainism, *ahiṃsā* (nonviolence) is the remedy for all miseries, sufferings, and cruelties of life. This *ahiṃsā* in its manifold aspects has found a full treatment in the *Ācārāṅga Sūtra*, the earliest text in the Jain canon. We are here told that all love life and hate death and suffering: "All beings are fond of life, they like survival; life is dear to all."[15] And further: "[T]here are beings who are blind, sunk in darkness; they experience ups and downs." "Attached to things sensual, they bewail bitterly, and on account of desires, fail to get emancipation." "Blinded and immersed in worldly pleasures, the fool with bondage unsevered and attachment not cut off, dwells in darkness, being ignorant, and is never able to get at the command." "Restrain yourself and you will be emancipated from suffering." "The violent person is one who is umindful of and addicted to worldly pleasures."[16]

These prophetic injunctions from the *Ācārāṅga Sūtra* are exclusively concerned with the observance of nonviolence. Another such injunction runs as follows:

> Thus do I say—The Revered Ones (*Arahamtas*), and the Lords (*bhagavantas*) of the past, present and future, all say thus, speak thus, enjoin thus, explain thus—all breathing, all existing, all living, all sentient creatures should not be killed, nor driven away. This is the discipline which is pure, eternal, inalterable, and declared by the enlightened ones who have comprehended the nature of the world.[17]

The *Ācārāṅga Sūtra*'s approach to the interrelationship of living beings is very much like that of the Upaniṣads, which also emphasize the interconnectedness of life. This is clear from the identification of the killer and killed in the following verse that explains the essential nature of *ahiṃsā*: "You are the one whom you intend to kill, you are the one you intend to tyrannize, you are the one whom you intend to torment, you are the one you intend to punish and drive away. The enlightened one who lives up to this dictum neither kills nor causes others to kill." Thus, "Bondage and emancipation are within yourself."[18]

Special mention is made in the *Ācārāṅga Sūtra* of the vegetable kingdom (trees and plants), to explain how they show signs of life that should be protected and respected:

> As the nature of this [person] is to be born and to grow old, so is the nature of that [plant] to be born and to grow old; as this has reason, so that has reason. . . ."[19]

> [A]s this falls sick when cut, so that falls sick when cut; as this needs food, so that needs food; as this will decay, so that will decay; as this is not eternal, so that is not eternal; as this takes increment, so that takes increment; as this is changing, so that is changing. One who injures these [plants] does not comprehend and renounce the sinful acts; one who does not injure these, comprehends and renounces the sinful acts. Knowing them, a wise person should not act sinfully towards plants, nor cause others to act so, nor allow others to act so. One who knows these causes of sin relating to plants is called a reward-knowing sage.[20]

The Jain, as a believer in the life of plants and particles of earth, water, air, and fire or energy, is very scrupulous about matters of diet. Jain vegetarianism is quite restrictive; one avoids the use of as many varieties of vegetables as possible. Vegetarianism is gaining ground among intellectuals who appreciate that alcohol and meat consumption entails the use of enormous amounts of grain to produce liquor and to feed cattle. François Peroux, director of the Institute of Mathematics and Economics in Paris, has suggested that if meat and alcohol consumption in the West were reduced by 50 percent, the grain that would become available would be enough to solve all hunger and malnutrition problems in the Third World. Deaths caused by automobile accidents and cardiovascular illness would also be reduced in the West if the consumption of liquor and meat were decreased.[21]

The Moral Code of Jainism and *Ahiṃsā* as Categorical Imperative

Jainism prescribes two types of vows. The great vows (*mahāvrata*) are to be practiced by members of the monastic orders and the small vows (*aṇuvrata*) by the laity. Abstinence from violence, falsehood, stealing, carnality, and possessiveness are the great vows. The small vows refer to a commitment to partial abstinence according to one's capacity, and the great vows refer to a commitment to complete abstinence, observed absolutely and without relaxation.

In principle, members of monastic orders are not involved in damaging the environment in any way. Because of the limited nature of their vows, only laypersons hold the responsibility for ecological problems caused by overpopulation, overcompetition, and overindulgence based on the wanton exploitation of nature. The small vows are prescribed in order to restrain laypeople from those tendencies that will lead to such problems. Among the small vows prescribed for laypersons are the following: refraining from movement beyond a limited area in order to avoid all harmful activities beyond that area; restricting movement to an even more limited area so as to expand the area of immunity from exploitative activities; refraining from wanton destruction of the environment by thought, word, or deed; and limiting the use of consumable and nonconsumable goods. Wanton destruction includes five varieties: evil thought of conquest, subjugation, killing, mutilating, hurting, and so on; evil counsel to torture animals and indulge in harmful activities; negligent conduct, such as recklessly cutting trees, digging in or flooding fields, or burning forests; applying lethal weapons; and malicious indoctrination. Limiting use of consumable and nonconsumable goods refers to keeping to a minimum one's food, drink, cosmetics, rich clothes and jewelry, beds, chairs, vehicles, and so on. In brief, Jain ethics insists on maximizing beneficial interaction and minimizing the harmful interference with nature.

Ahiṃsā in Jainism is a universal law, a rational maxim designed to govern all of one's actions. It is thus a categorical imperative as explained in Kant's *Critique of Practical Reason*. As an illustration of such a universal law with reference to the moral principle of forgiveness, I would like to quote the following from *That Which Is*:

> Forgiveness depends upon controlling anger and practicing tolerance in adverse situations. . . . [I]t requires forbearance and gratitude that worse has not happened. To practice unconditional forgiveness, we should see ourselves as the source of anger. . . . [A]ngry abuse from another should be countered by looking to oneself for the cause of the anger. If the cause can be found within oneself the other should be forgiven for [t]his anger. Even if the fault does not lie with oneself, the other should be forgiven because [t]his anger is due to ignorance. The ignorant should always be forgiven. If someone accuses us covertly, [that person] should be forgiven because he [or she] did not resort to physical violence. If [one] did resort to beating, that person should be forgiven because he [or she] did not kill us. If [one] did attempt to kill us, [that person] should be forgiven for not distracting us from the religious path. We should always find reason to forgive a person who harms us and should remember that whatever misfortunes confront us, they are due to our past karma.[22]

This text specifies a comprehensive approach to the cultivation of nonviolence through the uprooting of all anger.

Jain Epistemology and Logic: Nonabsolutism and Standpoints in Logic

Knowledge is an intrinsic property of the soul according to Jainism. But this knowledge lies covered by the veil of ignorance. The cover, however, has "holes" through which the soul has knowledge of the world it lives in. The holes are apertures that differ in capacities from soul to soul and also from circumstance to circumstance. The soul consequently cannot get an absolutely true knowledge of the things it attempts to know. The apertures distort the vision, as it were, to make it useful to the biological and physical needs of the soul. Each soul has its own view of the world, which differs from the views of other souls. No view is perfect, though each view has its own utility. Thus the world is not as it seems to be, and therefore it is not untrue to think that, if we shut our eyes, the objects we had been seeing do not remain as they appeared to us.

The Jain philosopher Haribhadra asserts that the biased thinker twists logic to suit his preconceived theory, whereas an unbiased truth seeker revises his theory to follow the dictum of logic:

> *āgrahī vata ninīsati yuktiṃ tatra*
> *yatra matirasya niviṣṭā*
> *pakṣapātarahitasya tu yuktir yatra*
> *tatra matireti niveśam.*[23]

Haribhadra here is in perfect agreement with the distinction Bertrand Russell makes between classical and modern logic:

> In classical logic [ancient Indian logic in the case of Haribhadra] hypotheses which seem *prima facie* possible are professedly proved impossible, and it is decreed in advance that reality must have a certain special character. In modern logic [Jain logic in the case of Haribhadra] on the contrary, while the *prima facie* hypotheses as a rule remain admissible, others, which only logic would have suggested, are added to our stock, and are very often found to be indispensable if a right analysis of the facts is to be attained. The old logic put thought in fetters, while the new logic gives it wings.[24]

The analysis of empirical experience and logic in Jainism clearly implies that no absolute philosophy is admissible, and there are standpoints in logic. Logic depends on science for its universal propositions, which are liable to variation with the progress in science. No absolute truth is thus possible, and logic has to depend on standpoints in science. Philosophy is becoming scientific through the simultaneous acquisition of new facts and logical methods.

Ahiṃsā in Practice

In the romantic dialogue in *Uttarādhyayana Sūtra* 22, Rathanemi makes amorous advances to Rājimati, who had become a nun, but fails in his mission. In the background of the dialogue, Prince Ariṣṭanemi, Rathanemi's younger brother, rides on the elephant of Vāsudeva under a raised umbrella, surrounded by an army in procession, coming to marry Princess Rājimati. On learning that animals were kept in cages and enclosures to be killed for the marriage party, he refuses to marry, saves the animals, and retires to Mount Girnār to lead the life of an ascetic. This fabled event was a milestone in the history of the protection of animals. Ariṣṭanemi, the former prince, became the twenty-second Tīrthaṅkara of the Jainas. The next Tīrthaṅkara, Pārśvanātha, finding a snake burning amid fuel lit by an ascetic who was surrounded by five fires, did not remain silent but raised his voice against this form of penance. Both these stories about the twenty-second and twenty-third Tīrthaṅkaras underscore the concern for animals and their place within nature. Included in the stories of the twenty-fourth Tīrthaṅkara is one of the ascetic Jayaghoṣa, a disciple of Mahāvīra, who ignores the risk of opposition and attends a sacrificial ritual to save the animal—and also the sacrificer, from his violent habits.

Among historical records of the practice of *ahiṃsā*, perhaps the best known are the inscriptions of the emperor Aśoka whose early faith, according to Edward Thomas, was Jainism.[25] According to Hendrik Kern, Aśoka's ordinances concerning the sparing of animal life agree much more closely with the ideas of the Jains than with those of the Buddhists.[26] Girnār Rock Edict No. 1 reads:

> This record relating to the Dharma has been caused to be written by King Priyadarśi, Beloved of the Gods.
>
> Here no living being should be slaughtered for sacrifice and no festive gathering should be held. For King Priyadarśi [Aśoka], Beloved of the Gods, sees manifold evil in festive gatherings. There is, however, one kind of festive gathering which is considered good by King Priyadarśi, Beloved of the Gods.
>
> Many hundred thousands of living beings were formerly slaughtered every day in the kitchen of Priyadarśi, Beloved of the Gods, for the sake of curry. But now, when this record relating to Dharma is written, only three living creatures are killed daily for the sake of curry, viz. two peacocks and one deer. Even this deer is also not slaughtered regularly. These three living beings too shall not be killed in the future.[27]

The pillar at Rampurva includes the following edict:

> Thus saith King Priyadarśi, Beloved of the Gods. Twenty-six years after my coronation, I have declared the following species of animals

exempt from slaughter, viz. parrots, mainas, ruddy geese, wild geese, *nandimukhas*, *gelatās*, bats, queen ants, terrapins, boneless fish, *vedaveyakas*, *gangapuputakas*, skate-fish, tortoises and porcupines, leaf-hares, twelve-antler stags, bulls set at liberty, household vermin, rhinoceroses, white pigeons, village pigeons and all quadrupeds which are neither useful nor edible.

Those she-goats, ewes and sows, which are either pregnant or milch, are not to be slaughtered nor their young ones which are less than six months old. Cocks are not to be caponed. Husks containing living beings should not be burnt. Forests must not be burnt either uselessly or in order to destroy living beings. The living must not be fed with the living.

At the three Chāturmāsīs and at the full-moon of the month of Tishya, for three days in each case, viz. the fourteenth and fifteenth of one fortnight and the first of the next and invariably on every fast day, fish is exempt from slaughter and should not be sold. And on some days, not only these but other species of beings should not be killed in the elephant-forests and the fisherman's preserves.

On the eighth of each fortnight and on the fourteenth and fifteenth, on the Tishya and Punarvasu days, on the three Chāturmāsī days and on every auspicious day, bulls are not to be castrated. And he-goats, rams, boars and such other animals as are usually castrated should not be castrated on those days. Horses and bullocks should not be branded on the Tishya and Punarvasu days, on the Chāturmāsīs and during the fortnights associated with the Chāturmāsīs.

Up to the time when I completed twenty-six years after my coronation, the release of prisoners has been ordered by me for twenty-five times during the period in question.[28]

Aśoka's grandson, King Samprati, stopped the slaughter of animals and ensured their protection.

During the early part of the second millenium, the great Śaiva king Siddharāja of Gurjar, in obedience to the advice of Maladhāri Ācaryā Abhayadeva and Ācaryā Hemacandra, gave protection to animals, beasts, and birds. His successor, Kumarapala, is famous for his declaration of nonslaughter (*amari*) throughout his kingdom. Ācaryā Jinaprabhasūri, at a later time, persuaded Muhammad Tughlak, the sultan of Delhi, to perform many acts of animal protection. Ācaryā Hīravijaya and other Jain monks, such as Shāntichandra and Bhānuchandra, convinced the great Mogul emperors Akbar, Jahangir, and others to legislate animal protection. Because these emperors had been so addicted to meat-eating and to hunting, these are certainly glorious instances of *ahiṃsā* practice.

The Jain monks and their lay followers courageously carried their mission to the mighty emperors who followed different faiths, and ultimately their efforts were crowned with success. The famous *farmans*, or special animal protection laws, of those emperors are still before us and testify to the dynamic character of the principle of nonviolence advocated by the Jains.[29]

In the modern world, various practices and institutions still exist throughout India that have been established by the Jain community for animal protection. There are thousands of *pinjrapoles* (animal refuges), organized and financed by Jains, in the towns and villages of Gujarat and Rajasthan, for the protection of helpless beasts and birds, whom even their owners forsake and who do not enjoy even the provision of drinking water. The Bombay Humanitarian League has worked tirelessly to stop animal slaughter at religious functions, as well as the custom of eating meat, working at both the individual and societal level. The league has saved thousands of animals from slaughter.

Perhaps the most famous proponent of nonviolence in the twentieth century was Mahatma Gandhi, the Hindu reformer, deeply influenced by Jainism, who combined love and nonviolence. Nonviolence, according to Gandhi, is the law of the human race and is infinitely greater than and superior to brute force.[30] Nonviolence "does not simply mean non-killing. *Himsa* means causing pain to or killing any life out of anger, or from a selfish purpose, or with the intention of injuring it. Refraining from so doing is *ahimsa*."[31] *Ahiṃsā* and truth (*satya*), according to Gandhi, "are so intertwined that it is practically impossible to disentangle and separate them. . . . Nevertheless, *ahimsa* is the means; Truth is the end. Means to be means must always be within our reach, and so *ahimsa* is our supreme duty."[32] "In its positive form, *ahimsa* means the largest love, greatest charity." According to Gandhi, "If I am a follower of *ahimsa*, I *must love* my enemy." This active form of *ahiṃsā* "necessarily includes truth and fearlessness. . . . A [person] cannot practice *ahimsa* and be a coward at the same time. The practice of *ahimsa* calls forth the greatest courage."[33] Gandhi brought this philosophy to the masses of India and revolutionized the South Asian world. Not only did Gandhi successfully gain India's independence from the British Empire, he also set the social agenda for a more egalitarian Indian society, working to combat the harm caused by caste and religious divisions. He also criticized the human tendency of taking beyond one's true need to satisfy one's greed, an insight with ecological wisdom before its time.

The Reasons for Ecological Crisis and the Remedy

The root of the ecological crisis is a spiritual crisis of self-centered greed, aided and abetted by ingenious technologies no longer properly restrained. Anger, pride, deceit, and greed have vitiated the human mind beyond imagination, and they are now horribly visible in nuclear explosions and smoldering rain forests and psychologically apparent in the rampant consumerism of our times. How can we survive on a planet of ten billion points of infinite greed and pride?

Pollution, extinction of species, and destruction of forests and wild life are crimes against the earth and against humanity. We have a moral obligation toward nonhuman creation. Rajotte writes: "The most urgent task of both science and religion is to assert the unity and sacredness of creation, and to reconsider the role of humans in it."[34] Traditional Jain teachings can serve to remind the world of the power and complexity of nature as well as the moral advantage of living simply according to the vows associated with nonviolence.

Lily de Silva has said: "We have to understand that pollution in the environment has been caused because there has been psychological pollution within ourselves. If we want a clean environment, we have to adopt a lifestyle that springs from a moral and spiritual dimension." We have to follow a simple, moderate lifestyle, she continues:

> eschewing both extremes of self-deprivation and self-indulgence. Satisfaction of basic human necessities, reduction of wants to the minimum, frugality and contentment are its important characteristics. Every individual has to order their life on moral principles, exercise self-control in the enjoyment of the senses, discharge their duties in their various social roles, and behave with wisdom and self-awareness in all activities. It is only when each person adopts a simple moderate lifestyle that humanity as a whole will stop polluting the environment. This seems to be the only way of overcoming the present ecocrisis. . . . With such a lifestyle, humanity will adopt a non-exploitative, non-aggressive, caring attitude towards nature. We can then live in harmony

> with nature, using its resources for the satisfaction of our basic needs. Just as the bee manufactures honey out of nectar [by taking from a large number of different flowers], so we should be able to find happiness and fulfillment in life without harming the natural world in which we live.[35]

Each faith has something to offer for meeting the ecological crisis. None of them, even Jainism itself, historically has faced the problems we face today. Still, each of them has enunciated principles that have relevance for present times. In many ways, Jainism, with its intricate theories about life and its practices designed to preserve and protect life, can provide interesting insights into how to cultivate a lifestyle friendly to environmental values.

Some Jain Guidelines to Meet the Ecological Crisis

In order to engage traditional Jain ethics in the current discussion regarding environmental degradation, in might be helpful to articulate Jain principles that can be applied to help cultivate ecologically friendly behavior.

1. Do not kill. Do not let others kill. Find whatever means possible to protect life.
2. Always speak truthfully and constructively. Resist injustice even if it may threaten your own safety.
3. Make every effort to reconcile and resolve all conflicts, big or small, peacefully and by consensus.
4. Do not make the accumulation of wealth an aim of your life. Live simply and share time, energy, and material resources with those who are in need.
5. There is no absolute truth. No doctrine, theory, or ideology is perfect.
6. Practice nonattachment to views. Remain open to receive others' viewpoints. Do not force others to adopt your views.
7. Do not lose awareness of the existence of suffering in the world. Find ways to be with those who are suffering.
8. Do not maintain anger and hatred. As soon as they arise, concentrate on your breathing to see their nature.

If Jains are able to let these ideas govern their behavior, then they will be taking an important step toward alleviating the greed and pride that lead to a disregard for one's environment. Observing nonviolence to all beings reduces hatred, as do truthfulness and reconciliation. Minimizing one's involvement with status and riches allows for appreciation of the ordinary joys of life. Flexibility in regard to the ideas of others helps reduce judgments and expectations. Being open to the suffering of others allows one to develop compassion. By strengthening themselves to resist the various temptations put forth by technology and a consumer mentality, Jains can perhaps provide an example for living lightly on the planet earth.

Notes

1. *Daśavaikālika Sūtra* 6.8.
2. Freda Rajotte with Elizabeth Breuilly, "What Is the Crisis?" in *Christianity and Ecology*, ed. Elizabeth Breuilly and Martin Palmer (London: Cassell Publishers, 1992), 8.
3. Umāsvāti, *Tathvārtha Sūtra; That Which Is*, trans. Nathmal Tatia (San Francisco: HarperCollins, 1994), 44, 197.
4. Freda Rajotte with Elizabeth Breuilly, "Treatment for the Earth's Sickness—The Church's Role," in *Christianity and Ecology*, ed. Breuilly and Palmer, 105.
5. Norman Solomon, "Judaism and the Environment," in *Judaism and Ecology*, ed. Aubrey Rose (London: Cassell Publishers, 1992), 27.
6. Ibid., 39, with a reference to Genesis 2:15.
7. Matthew 5:39, 44–46 New English Bible.
8. Matthew 6:25 New English Bible.
9. Qur'ān 24:45, quoted in Al-Hafiz B. A. Masri, "Islam and Ecology," in *Islam and Ecology*, ed. Fazlun Khalid with Joanne O'Brien (London: Cassell Publishers, 1992), 10.
10. Ibid., 20.
11. Ibid., 18.
12. The Hindu philosophy of divine incarnations, on the other hand, assigns to God the task of universal redemption, making this aspect of Hinduism similar to the three prophetic monotheisms mentioned in the above section.
13. Umāsvāti, *Tathvārtha Sūtra; That Which Is*, trans. Tatia, 5.21, 6.3, and 7.1.
14. "Even the Stones Smile: Selections from the Scriptures," ed. Martine Batchelor, in *Buddhism and Ecology*, ed. Martine Batchelor and Kerry Brown (London: Cassell Publishers, 1992), 4.
15. *Ācārāṅga Sūtra* or *Ayaro* (New Delhi: Today and Tomorrow's Printers and Publishers, 1981), 1.2, 3.63.
16. Ibid., 1.6.1, 9–10; 1.6.1.7; 1.4.4.45; 1.2.2.64; 1.4.68–69.
17. Ibid., 1.4.1.1.
18. Ibid., 1.1.5.101–2; 3.1.5.2.36.
19. "The plants know the seasons, for they sprout at the proper time, the Asoka buds and blossoms when touched by the foot of a well attired girl, and the Vakula when watered with wine; the seed grows always upwards: all this would not happen if the plants had no knowledge of the circumstances about them. Such is the reasoning of the commentators." *Jaina Sutras*, trans. Hermann Jacobi, vol. 22 of *Sacred Books of the East,* ed. F. Max Müller (Delhi: Motilal Banarsidass, 1964), 10.
20. *Ācārāṅga Sūtra* 1.1.5.112–17.
21. Thich Nhat Hanh, *Interbeing, Fourteen Guidelines for Engaged Buddhism*, ed. Fred Eppsteiner (Delhi, 1997), 43.
22. Umāsvāti, *Tathvārtha Sūtra; That Which Is*, trans. Tatia, 221.
23. Unfortunately, Dr. Tatia passed away before we were able to obtain a citation for this quotation. *Ed.*
24. Bertrand Russell, *Our Knowledge of the Eternal World* (London: Allen and Unwin, 1952), 68.

25. Edward Thomas, *Jainism, or, the Early Faith of Asoka* (London: Trubner and Co., 1877), 30–32.

26. *Indian Antiquary*, 5:275.

27. *Inscriptions of Asoka*, trans. D. C. Sircar (New Delhi: Publications Division, Government of India, 1967), 39–40.

28. Ibid., 71–72.

29. Pandit Sukhalji, *The World Pacifist Meeting and the Role of Jainism* (Calcutta: Jain Reception Committee, n.d.), 14.

30. *Selections from Gandhi*, ed. Nirmal Kumar Bose (Ahmedabad: Navajivan Publishing House, 1957).

31. Ibid., 155.

32. Ibid., 13–14.

33. Ibid., 157–58.

34. Rajotte with Breuilly, "Treatment for the Earth's Sickness," 104.

35. Lily de Silva, "The Hills Wherein My Soul Delights: Exploring the Stories and Teachings," in *Buddhism and Ecology*, ed. Batchelor and Brown, 29.

Jain Ecological Perspectives

JOHN M. KOLLER

At a time in global history when violence toward human, plant, and animal life—and toward the very environment that makes the various life-forms possible—threatens to destroy whole societies, human life on this planet, and, perhaps, even the life of the planet itself, it is imperative that we seek nonviolent solutions to our problems. From a Jain perspective, the threat to life that we face arises from a faulty epistemology and metaphysics as much as from a faulty ethics. The moral failure to respect the life of others, including life-forms other than human, is rooted in dogmatic, but mistaken, knowledge claims that fail to recognize other legitimate perspectives. Because existence itself is complex, subtle, and many-sided (*anekānta*), unless the knowledge that grounds action reflects this many-sidedness of reality, it will produce actions that are destructive of existence opposed to the interests of the agent's own limited and biased perspective. As Umāsvāti says: "A person with a deluded world-view is like an insane person who follows arbitrary whims and cannot distinguish true from false."[1]

That is why the underlying philosophical question is, from a Jain perspective, "How are we to avoid the destructive violence that results from courses of action rooted in one-sided ideological dogmatism?" Because the ideological dogmatism underlying violence is grounded in epistemological claims that, though limited and only partially true, are mistaken for absolute truths, the answer, in part, is to find an alternative epistemology, an epistemology that can support dialogue and negotiation among diverse perspectives and claims. Such an epistemology, a perspectival and inclusive epistemology, I suggest, might be found in the Jain philosophy of nonabsolutism (*anekānta-vāda*).

What I am suggesting is that we view the Jain metaphysics of nonabsolutism (*anekāntavāda*), supported by the epistemological theory of viewpoints (*nayavāda*) and the sevenfold scheme (*saptabhaṅgi*) of qualified predication (*syādvāda*), as providing a basis for the central moral principle of *ahiṃsā*. Because violence proceeds from intolerance rooted in ideological absolutism, *ahiṃsā* requires a firm foundation for tolerance, a foundation provided by the philosophy of *anekānta*. The *anekāntavāda* philosophy can be seen as providing an ontological basis for the principle of nonviolence. It also grounds an epistemological basis for the respect for others that *ahiṃsā* incorporates. The perspectivalism embodied in the theory of epistemic standpoints (*nayavāda*) recognizes that ordinary, nonomniscient, knowledge claims are always limited by the particular standpoint on which they are based. Consequently, claims from one perspective must always be balanced and complemented by claims from other perspectives. This has important ecological implications, for it legitimizes considerations from nonhuman perspectives, enabling us to consider the effects of our actions on nonhuman life-forms and environments.

From a metaphysical perspective, Jainism can be viewed as transforming the principle of *ahiṃsā*, embodied in the respect for the life of others, into epistemological respect for the views of others, thereby establishing a basis for reconciling conflicting ideological claims. What does "epistemological respect for the views of others" mean? This question has been discussed in some of the recent literature on *anekāntavāda*. John Cort, for example, in a paper in *Philosophy East and West*, argues against a certain view that intellectual nonviolence is implicit in *anekāntavāda*, a view he calls "intellectual *ahiṃsā*."[2] His argument is two-pronged. The first prong attempts to show that *anekāntavāda* was developed to maintain the Jain view of the soul (*jīva*) as both eternal and changing. Here, I think, Cort is correct, as can be shown, at least in part, by tracing the development of *anekāntavāda* out of the earlier method of analysis and resolution called *vibhajyavāda*, as I have done in a forthcoming paper entitled "*Avyākata* and *Vibhajya* in Early Buddhism and Jainism."[3]

The second prong of Cort's argument attempts to show that Jain intellectuals were by no means completely tolerant of other views. In the first place, they recognized only their own view as fully correct; all other views were, at best, only partially correct. Second, they vigorously attacked views they regarded as wrong and pernicious, often ridiculing these views. This leads Cort to claim that Jain thinkers "did not end up in a position implied by H. R. Kapadia and others as the position of "intellectual *ahiṃsā*." The reason he gives is that "They did not advance a position of total relativity, nor did they lack standards by which to argue that their own *darśana* was not only distinct from but in their view better than the other *darśanas*."[4] If practicing "intellectual *ahiṃsā*" means that Jain thinkers regarded all views as equally valid, then it is clear that most Jain scholars did not do so. But it would hardly be reasonable to expect them to concede that all arguments and all views are equal, or that other views were equal or superior to their own. After all, they were committed to the truth of the Jain view, and, as scholars, were committed to explaining and defending their view. Furthermore, when the *anekāntavāda syādvāda* epistemological methodology is seen as a continuation and refinement of the *vibhajya* method, it becomes clear, as Jayandra Soni points out, that *anekāntavāda* is not a theory of indeterminacy or epistemological relativism.[5] It then also becomes clear that the view of *anekāntavāda* that W. J. Johnson rejects, the view "that *anekāntavāda*'s exclusive function is to promote nonviolence at the intellectual level,"[6] is a mistaken view, a view not held by any traditional Jain thinkers so far as we know.[7] But this does not alter the fact that both the *nayavāda* and the *syādavāda*, because they explicitly recognize the multiplicity of perspectives from which something can be viewed and the limitations of claims made from a particular perspective, lend themselves to a respectful acknowledgment of the partial truth of an opponent's views.

A less extreme version of intellectual *ahiṃsā* would allow Jain thinkers to maintain the correctness of their own view, to recognize the inferiority of other views, and to criticize other views in terms of their weaknesses, but to do so respectfully, recognizing their partial correctness. This would be a middle way between absolutism and relativism, allowing, in the words of Christopher Chapple, "The Jaina outlook toward the ideas of others [that] combines tolerance with a certainty in and commitment to Jaina cosmological and ethical views."[8]

In qualifying his argument against the strong form of intellectual *ahiṃsā* claimed by A. B. Dhruva[9] and H. R. Kapadia,[10] Cort comes close to recognizing this less extreme form of intellectual *ahiṃsā*, noting the respect that various Jain thinkers showed for the views and arguments of their rivals. He notes Paul Dundas's comment that Haribhadra "at times [showed] remarkable willingness to evaluate rival intellectual systems on the basis of their logical coherence alone."[11] Cort also approvingly quotes the famous statement from Haribhadra's *Lokatattvanirṇaya* cited by Dundas: "I do not have any partiality for Mahavira, nor do I revile peoples such as Kapila [founder of the Hindu Saṃkhya system]. One should instead have confidence in the person whose statements are in accord with reason."[12]

This strong statement of respect for the views of others by Haribhadra is rendered suspect in the eyes of some scholars because the biographies of Haribhadra present a rather different picture, a picture of a man openly hostile to his opponents, who in some accounts murdered hundreds of Buddhists.[13] But as Phyllis Granoff points out, the biographies, compiled centuries after Haribhadra's death, are at odds with the picture of Haribhadra that emerges from his own writings. In his writings Haribhadra "is clearly a man of religious tolerance, of quiet respect for differences and particularly of respect for Buddhist ethics and spiritual practices. . . ."[14] Why the biographies contain a distorted picture of Haribhadra's tolerance of and respect for the views of his opponents is unclear, but there seems little doubt that in his writings he strongly exemplified respect for his opponents' arguments and views, even while attempting to show how and why they were wrong.

From an espistemological perspective, the use of *syāt* as a logical operator that transforms absolute claims into qualified claims, according to *syādvāda*, particularly as operationalized through *nayavāda*, provided a way to achieve the nondogmatic epistemology required to put *ahiṃsā* into action.

Used in the service of *anekāntavāda* as a method of reconciliation, the *nayavāda* has the poential to eliminate violent argument between ideological opponents by methodically both disarming and persuading them. Operationalized in the sevenfold predicational formula (*saptabhaṅgi*) of *syādvāda*, the *nayavāda* can be seen as a sophisticated method of partially conceding the opponent's thesis in order to avert his argument, while at the same time persuading him to consider the other side of the case. It is thus a method of reconciling opposites and avoiding violence, making it an attractive basis for ecological thought and practice.

What are the *naya*s, and how does the theory of *naya*s provide an epistemological basis for nonviolence and ecological thought? Nonviolence is understood in Jainism as *ahiṃsā*, the basis of all morality. Jainism embraces a very strict and far-reaching concept of *ahiṃsā*. In analyzing a section of the *Śrāvakaprajñapti* that deals with a debate over the role of intentionality in acts of violence, Granoff points out that, whereas the Buddhists claim that unless a person intended the violence that follows an act the person is not guilty of performing a violent act, the Jains claim that if an act produces violence, then that person is guilty of committing a violent act even if the violence was not intended.[15] The example given is that if a monk unknowingly offers poisoned food to his brethren and they die from the poisoned food, according to the Jain view the monk would be guilty of performing a violent act, but according to the Buddhist view the monk would not be guilty. The crucial difference between the two views is that the Buddhist view excuses the act, categorizing it as nonintentional because the monk did not know that the food was poisoned, whereas the Jain view regards the act as intentional because the monk is responsible for his ignorance, and therefore for any act that follows from this ignorance. As Granoff says with respect to this example: "The Jains argue that all violence is intended violence; they argue that it is not possible for a person to be so ignorant and yet not guilty. His very ignorance and carelessness constitute an intent to do violence and imply correspondingly his guilt."[16]

According to Granoff, "One of the most informative discussions between a Buddhist opponent and the Jains on the issue of intentionality is the commentary of Śīlāṅka to the *Suyagaḍaṅga*. . . ." The example that Śīlāṅka discusses is that of a person who roasts a child who is covered up on a fire, thinking the child to be a gourd. Because the person had no intention to commit murder and was ignorant of what he did, he did not commit murder according to the Buddhist analysis. Śīlāṅka, on the other hand, argues that the man did, indeed, commit murder. The argument is that it is this very ignorance that results in bondage, and therefore it cannot be argued that ignorance excuses bad deeds. If only good intentions counted, then, for example, killing a person in order to release him from his bad *karma* would be a blameless act, a position that both Jains and Buddhists explicitly reject.[17] Thus, according to Jainism the moral imperative to practice *ahiṃsā* includes the requirement to remove the ignorance that would prevent a person from seeing the violence embodied in his or her actions.

The term "*ahiṃsā*," like "nonviolence," is negative, but the principle is entirely positive. It grows out of a philosophy that recognizes the community of all living organisms and that sees love as the basis of relationship between all the members of this community, not merely human community. *Ahiṃsā* embodies the realization that all life belongs to the same global family and that to hurt others is to destroy the community of life, the basis of all sacredness. Thus, *ahiṃsā* requires not only that we avoid hurting other life-forms, but that we must endeavor to help each other.[18] Indeed, Umāsvāti, the great Jain teacher of the second century, defines the purpose of life-forms as helping each other: "Souls [*jīvas*] exist to provide service to each other."[19] Because *ahiṃsā* is really a principle of respect for all life-forms, it is a fundamental ecological principle, an ecological principle that is grounded both in the understanding that all living things, despite their differences, embody the same kinds of *jīvas*, or life-principles, and in the understanding that these *jīvas* cycle through a vast number of bodies and environments. For these two reasons all living things are entitled to the same respect, to nonviolation of their bodies and environments.

The *nayas*, or standpoints, may be thought of as different points of view taken by someone searching for the truth. According to Akalaṅka, in the *Sanmati Tarka*, the standpoints are the presuppositions of inquirers, embodying the points of view from which they are investigating the thing in question.[20] In ordinary cognition, as opposed to omniscient cognition, the knower necessarily sees the thing from a particular point of view. Consequently, the nature of the thing that is revealed to him is necessarily conditioned and limited by this particular point of view, enabling him to have only partial, incomplete knowledge of it. As Siddhasena says: "Since a thing has manifold character, it is [fully] comprehended (only) by the omniscient. But a thing becomes the subject matter of a *naya*, when it is conceived from one particular standpoint."[21] Thus, the *nayas* serve to categorize the different points of view from which reality might be investigated. Recognition of the fact that her inquiry is from only one of these standpoints enables the investigator to recognize the partial, limited nature of her knowledge, preventing her from becoming one-sided (*ekānta*) and dogmatic in her claims. It also encourages investigators to assume other perspectives, including the ecologically important perspective of the other as a persisting, but constantly changing, entity entitled to the same respect for life and happiness as oneself. For example, when one assumes the perspectives of other life-forms, such as animals or plants, it is possible to see and feel their connectedness to us and to feel their suffering when they are injured. Knowing how much like us they are and knowing that they are as dependent on their environment as we are, we have incentive to not injure them and to not destroy their environment.

With regard to the number and character of standpoints from which something may be investigated, it is generally agreed that although theoretically there are an unlimited number of them, two opposing standpoints are fundamental. On the one hand, things can be viewed in terms of their substantial being, emphasizing their self-identity, permanence, and essential nature. This standpoint regards sameness as fundamental. As an extreme view, it is exemplified by the Advaita teaching that Brahman alone is truly real. On the other hand, things can be viewed in terms of process, emphasizing the changes that they undergo. This standpoint emphasizes difference. In its extreme form it is exemplified by the Buddhist teaching of *pratītya samutpāda* as the nature of existence, a teaching that insists that everything is selfless (*anātman*) and impermanent (*anitya*).

When the differences within each of the two fundamental standpoints of sameness and difference are taken into account, we get the standard set of seven *nayas*, namely: *naigama* (the ordinary, or undifferentiated); *saṃgraha* (the general); *vyavahāra* (the practical); *ṛju-sūtra* (the clearly manifest); *śabda* (the verbal); *samabhirūḍha* (the subtle); and *evambhūta* (the "thus-happened"). The first three, *naigama*, *saṃgraha*, and *vyavahāra*, are standpoints from which to investigate the thing itself, as a substance, whereas the remaining four are standpoints from which to investigate the modifications that things undergo.

The Seven *Nayas*

In what follows I describe each of these seven basic epistemic standpoints, illustrating their differences by noting how, when we look at an ocean ecologically from the different *nayas*, we might see vastly different things. Although the examples of what is seen when the oceans are perceived from the different *nayas* are intended primarily to illustrate the differences between these different standpoints, they are also suggestive of the power of the *naya*s as a basis of ecological thought.

It seems quite appropriate that a paper presented at a conference on Jainism and ecology in 1998, the year designated by the United Nations as the Year of the Ocean, should at least suggest some of the kinds of contributions Jainism might make to a vitally important, but sadly neglected, environmental issue. As Elisabeth Mann Borgese points out in her book *The Oceanic Circle*, the ocean is an essential part of the biosphere, a crucial factor in the carbon cycle, and a major determinant of the planet's climate.[22] Not only does the ocean cover almost three times as much of the planet's surface as the land does, but it contributes an even larger percentage of the planet's ecosystem services. All of the forms of life on our planet that environmentalists are concerned to protect—and that *ahiṃsā* admonishes us not to hurt—began in the ocean and, in a variety of ways, continue to depend on the ocean for their life. The various landmasses, including the large continents, are islands floating and drifting on the ocean, which connects and supports them, providing the environment not only for all the forms of life that live in the water, but for those that live on the land as well. Three-fourths of the planet's oxygen is produced by marine plants, and more than 90 percent of the planet's carbon dioxide is contained in the ocean. Clearly, the ocean is the key to life on this planet, and an environmental philosophy or ecological view that ignores the ocean is seriously flawed. As Mann Borgese says: "It is clear by now to any . . . who want to see that we are killing ourselves by mishandling our environment, of which the ocean is the most vulnerable component."[23]

1. The *naigama* standpoint looks at a thing without differentiating between its substance and its qualities, not distinguishing between what it is and how it changes. It is a standpoint that takes the being of something to be more fundamental than its becoming, and thus does not differentiate among its various qualities and modes. From the *naigama* perspective, the ocean might be seen as constituting the major portion of the planet's surface. Its waters might be seen as the waters of life, primordial source and home of all life-forms. From this perspective the land is seen as simply a continuation of the sea, rather than seeing the seafloor as a continuation of the land. This is a perspective that would put the ocean at the center of thought, a perspec-

tive from which one could see that all protection of life must include protection of the ocean. Although a legitimate perspective, the *naigama* is lacking in specificity and therefore limited. The *nayavāda* reminds us of this, pointing out that it is only one among many perspectives and therefore needs to be balanced by integrating this perspective with other perspectives.

2. The *saṃgraha* standpoint emphasizes the generic character of a thing. The literal meaning of the word *saṃgraha*, "collection," furnishes an important clue, for different things—humans, animals, and minerals—can be collected or gathered together in the general category of "thing." At an even higher level of generality, all existing things can be referred to as *sat*, the existent. Thus, the *saṃgraha* standpoint ignores the differences between things in order to emphasize their sameness. From the *saṃgraha* perspective, the ocean is simply a large body of water, because this perspective ignores both the various life-forms in the ocean and particular transformations. But it is a perspective that enables us to see the ocean as a whole and to recognize both its place within a larger environment, for example, that of the life of the whole planet and the place of land within the ocean's environment. Most importantly, it is a perspective that enables us to recognize that the ocean is valuable in itself, and not merely valuable because of its usefulness for others.

3. The *vyavahāra*, or practical, standpoint has two related functions. On the one hand, it is the standpoint that enables one to function in a practical way, ignoring unnecessarily subtle or irrelevant complexities in accomplishing the task at hand. On the other hand, *vyavahāra*, as a standpoint that complements the *saṃgraha*, analyzes the differences among the different things collected into a genus, emphasizing their species characteristics. It is practically useful to know whether a thing is living or inanimate, and, if living, whether it is animal or plant, and so on. Thus, while *vyavahāra* is still a standpoint that emphasizes the essential and permanent characteristics of a thing, it is practical in the sense that it focuses on specific characteristics. From the *vyavahāra* perspective, we might see the ocean in terms of its economic aspects, as a medium of transportation, as a source of commercial food, of minerals. Or, also from a practical perspective, we might see it as a source of oxygen for life on earth, a part of the ecological balance of the planet. And these two perspectives might well be at odds with each other, generating conflict, for example, be-

tween those interests concerned to preserve the oceanic environment and those interests concerned to mine the ocean floors to obtain minerals at a minimum cost despite polluting the environment and disrupting the ocean's ecological balance. Other perspectives may be needed to resolve these conflicts.

4. The *ṛjusūtra* (literally, the "straight-thread") standpoint focuses on the momentary event that is clearly manifest in our present experience, directing our attention to the process of change occurring right now, in this very place. It ignores the underlying, enduring substance, focusing only on the fluctuating moment of change. Taken to the extreme, this standpoint sees only change, resulting in a view that denies being in favor of becoming. From this pinpoint perspective of the immediate present, the *ṛjusūtra naya*, because we are focusing only on the present actuality, we would not recognize the ocean's commercial potential or its historical role in creating the earth's environment, but we could see its variety of marine life-forms, its production of oxygen, and its functions in creating the earth's atmosphere. This perspective would also allow us to assess the present condition of ocean life.

5. The *śabda*, or verbal, standpoint investigates things from the standpoint of language, using the clues provided by grammar, semantics and syntax. Of particular interest are linguistic functions of tense and inflection, which direct our attention to the changing aspect of things. From the *śabda* perspective, we might see that the ocean cannot be described merely as water, but that because of its incredible diversity, many terms must be employed to describe it accurately. From this we could come to understand that the ocean is not simply a body of water, but that its mineral and plant life constitute an important part of its existence. This perspective might lead us to see the ocean—and ourselves—in terms of the myths and legends of the seas that have been created by many cultures—myths and legends that see water as the source and power of all life.[24]

6. The *samabhirūḍha*, or "subtle," standpoint looks at things from a linguistic perspective, focusing on the subtle distinction between the connotation and the denotation of words. For example, words like *rājan*, *nṛpa*, and *bhūpa* denote or refer to the same thing, the king, but each has different meaning, as seen from its different etymological formation. Thus, the meaning of *rājan* is "one who wears the royal

insignia"; *nṛpa* means "one who protects men"; and *bhūpa* means "one who protects the earth." Although all three terms refer to the king, they have quite different meanings, suggesting a world of difference beyond a common, shared reference. From the *samabhirūḍha naya*, we would investigate the meaning of various words that refer to the ocean, such as "ocean," "sea," "*sāgara*," noticing that different words and different etymologies suggest different aspects of the ocean to us. This would increase our awareness of the complexities of ocean life. This perspective might also help us distinguish between physical, cultural, economic, and political understandings of the ocean, thereby generating a better understanding of the whole.

7. The *evambhūta* (literally, "thus-happened") standpoint extends the *samabhiruḍha* standpoint even further, emphasizing the uniqueness of each speech event. Because each occurring event is unique, each speech act referring to the event is also unique, and the words uttered take on a unique meaning in reference to each particular event. It restricts the meaning of a particular word to its particular use at this moment. From the *evambhūta naya*, we would focus on the constantly changing modalities of the ocean, noticing that although it endures through time, its endurance is that of something constantly changing. Here we might notice the various causal factors at work in the unceasing transformations that constitute the life of the ocean, and we might begin to worry that dumping pollutants into the ocean or mining the ocean floor might produce changes that could radically change, and perhaps destroy, the ocean as we know it. This perspective might enable us to focus on the planet's kryosphere, its frozen fresh water that generates tens of thousands of icebergs each year, enabling us to see the vertical circulation of water from surface to the ocean's depth and back to the surface as icebergs form and melt. This in turn would enable us to understand the interactions between ocean floor, water columns, and atmosphere and to see distribution patterns of pollutants.[25]

Thus, we see that each *naya*, or standpoint, allows the investigator only a partial and therefore limited view of the object in question. The principal value of recognizing that a *naya* affords only a partial view of the object is that it enables one to distinguish between the limited view that results from a *naya* and the genuine knowledge that a valid means of knowledge, a *pramāṇa*, provides. This distinction, in turn, makes it possible to recognize when knowledge claims are excessive or one-sided (*ekānta*) because they confuse a *naya* with a *pramāṇa*. As one perceives the object from a combination of standpoints, one comes closer to seeing the object as it really is. But only by seeing it from all standpoints would one actually attain the kind of valid cognition that *pramāṇa*s alone can provide.

Nayavāda, in avoiding the one-sided errors of identifying existence with either the permanence and sameness of being, on the one hand, or with the ever-changing process of becoming, on the other, supports the metaphysical doctrine of *anekāntavāda* as a way of thinking about existence as simultaneously both being and becoming. *Anekāntavāda* accounts for things having the apparently exclusive characteristics of being and becoming (nonbeing) by recognizing that whatever exists is simultaneously of the nature of *dravya*, persisting substance, and *dharma*s, its changing features, which include its *guṇa*s and *paryāya*s. Thus, Umāsvāti defines the existent as substance, that is, unchanging, but then declares that it also has the triple character of originating, decaying, and persisting, as its various characteristics change.[26] But since becoming is the negation, the "is-not" of being, and since being is the negation, the "is-not" of becoming, Jain logic insisted on the middle ground between the extremes of "is" and "is not" in order to predicate both being and becoming of the same existent. Maintaining this middle ground led to the Jain development of *syadvāda*, a theory of predication that recognizes not only the predicates "is" and "is not," but also the predicate "inexpressible," a predicate that combines "is" and "is not." Combining the theory of standpoints with the three predicates leads to the famous sevenfold template for expressing important claims. If, for example, following our earlier ecological examples, we want to make claims about the ocean, we can make them in the following ways:

1. In a certain respect (from the general standpoint), the ocean is certainly a single body of water that persists through time. From this perspective the ocean will always be there, and its continuing existence is taken for granted.

2. But in another respect (from the standpoint of momentary process), the ocean is not a single body of water whose existence through time can be taken for granted, for it is actually an incredibly diverse complex of ever-changing processes, interlinked and mutually dependent, never the same, and constantly at risk.

3. In yet another respect (combining different standpoints), certainly both 1 and 2 are true, for while the ocean is an enduring entity, it is also a constantly changing entity.

Understanding that everything can be known in various ways, from various perspectives, leads naturally to a more balanced view, providing an understanding of the richness and complexity of existence, an understanding that encompasses the insight that other beings are not other to themselves; that they are themselves just as much as we are ourselves. It is this insight that enables us to see the other on its own terms, from its own side, rather than as merely the other. And this ability to see the other as no longer the other, but as identical to our own self, underlies the capacity for empathy and sympathy with the other that operationalizes *ahiṃsā*.

When we combine the underlying insight of *ahiṃsā* with the Jain view that bodies are constantly changing, breaking up, and transforming into other bodies, and that *jīva*s cycle through countless bodies before attaining liberation in the final release from a human body, we arrive at a profound basis for ecological thought.

It is because of this recycling of *jīva*s that the goal of liberation dominates Jain life, and that the practice of *ahiṃsā* is recognized as a condition for liberation, for karmic bondage means not only the relatively pleasant and blessed life of a human being, but also the untold lifetimes spent embodied as lower forms of life. The Jain view of life is essentially ecological because all forms of life are valuable; lower and higher forms constitute a single large community in which every life-form is entitled to respect and ethical treatment. The path to liberation is a path through this community of life, and requires the highest ethical standards extended to all life-forms. This is why Umāsvāti introduces his classic work explaining Jain philosophy with the words: "The enlightened worldview, enlightened knowledge, and enlightened conduct are the path to liberation."[27] According to Jain teachings, every soul cycles through tens of thousands of incarnations, ranging from fire, mineral, air, or vegetable bodies to those of plants, animals, humans, and gods. The ability to feel the intense sufferings of the soul in these various embodiments is powerfully revealed in an eloquent speech of the young prince, Mṛgapūtra, in which he begs his parents to allow him to leave home and take up the religious life in order to cut the bonds of suffering. A brief excerpt from his speech will suggest the importance and urgency of taking an ecological perspective in thought and action.

> From clubs and knives, from stakes and maces, from broken limbs,
> have I hopelessly suffered on countless occasions.
> By sharpened razors and knives and spears have I these many times
> been drawn and quartered, torn apart and skinned.
> As a deer held helpless in snares and traps,
> I have often been bound and fastened and even killed.
> As a helpless fish I have been caught with hooks and nets,
> scaled and scraped, split and gutted, and killed a million times. . . .
> Born a tree, I have been felled and stripped, cut with axes and chisels
> and sawed into planks innumerable times.
> Embodied in iron, I have been subjected to the hammer and tongs
> innumerable times, struck and beaten, split and filed. . . .
> Ever trembling in fear; in pain and suffering always,
> I have felt the most excruciating sorrow and agony.
> (*Uttarādhyayana Sūtra*, 19.61–74)

Mṛgapūtra's speech is a powerful example of the awareness and compassion aroused by the ability to see and feel things from the perspective of the other, whether the other is person, fish, or plant. It is precisely this ability to assume the standpoint of the other that arouses the energy and will to not harm the other, but to act so as to contribute to its well-being, that is at the heart of *ahiṃsā*. And it is this ability to assume the perspective of the other that makes it possible to see other life-forms, indeed, all of nature, as alive, as endowed, like us, with awareness and feelings, that enables the Jain principles of *anekānta-vāda* and *ahiṃsā* to work together as an effective basis for ecological thought.

Notes

1. Umāsvāti, *Tattvārtha Sūtra*, 1.33; as translated by Nathmal Tatia in *That Which Is* (San Francisco: HarperCollins Publishers, 1994), 23.

2. John E. Cort, "'Intellectual *Ahiṃsā*' Revisited: Jain Tolerance and Intolerance of Others," *Philosophy East and West* 50, no. 3 (2000): 324–47.

3. Forthcoming in *Proceedings from the International Conference on Jainism and Early Buddhism, Lund University, June 4–7, 1998*.

4. Cort, "'Intellectual *Ahiṃsā*' Revisited."

5. Jayandra Soni, "Philosophical Significance of the Jaina Theory of Manifoldness," in *Studien Zu-Interkulturellen Philosophie*, vol. 7, p. 285.

6. W. J. Johnson, "The Religious Function of Jaina Philosophy: *Anekāntavāda* Reconsidered," *Religion* 25, no.1 (1995): 41.

7. Soni, "Philosophical Significance of the Jaina Theory of Manifoldness," after arguing that *anekāntavāda* is a method of philosophical analysis and resolution of rival claims, notes that "For the Jaina thinkers *anekāntavāda* seemed to have been such a philosophical methodological tool and they do not seem to have themselves interpreted the method as being grounded on intellectual non-violence" (283).

8. Christopher Key Chapple, *Nonviolence to Animals, Earth, and Self in Asian Traditions*, SUNY Series in Religious Studies, ed. Harold Coward (Albany: State University of New York Press, 1993), 85.

9. *Malliṣeṇa, Syada-manjari*, ed. A. B. Dhruva (Bombay: Bombay University, 1933), lxxiv.

10. *Haribhadra: Anekāntajayapatāka*, ed. H. R. Kapadia, vol. 2 (Baroda: Oriental Institute, 1947), cxiv.

11. Paul Dundas, *The Jains* (London and New York: Routledge, 1992), 196.

12. Ibid., 197.

13. See Phyllis Granoff, "Jain Lives of Haribhadra," *Journal of Indian Philosophy* 17 (1989): 105–28.

14. Ibid., 106.

15. E. Phyllis Granoff, "The Violence of Non-Violence: A Study of Some Jain Responses to Non-Jain Religious Practices," *Journal of the International Buddhist Studies Association* 15, no. 1 (1992): 32.

16. Ibid.

17. Ibid., 33–34.

18. See John M. Koller and Patricia Koller, *Asian Philosophies* (Upper Saddle River, N.J.: Prentice Hall, 1998), 37.

19. *Tattvārtha Sūtra* 5.21.

20. Akalaṅka, *Sanmati Tarka* 3.47, ed. S. Sanhhavi and B. Doshi (Ahmedabad: Gujarat Paratattva Mandira Granthavali, 1924–31).

21. Siddhasena, *Nyāyāvatara*, 29, ed. A. N. Upadhye (Bombay: Jaina Sahitya Vikasa Mandala, 1971).

22. Elisabeth Mann Borgese, *The Oceanic Circle: Governing the Seas as a Global Resource* (Tokyo: United Nations University Press, 1998). See especially pp. 3–38.

23. Ibid., 35.

24. See, for example, the wonderful and powerful myths and stories collected in M. Morven, *Legends from the Sea* (New York: Crescent Books, 1980).

25. See Gotthilf Hempel, "The Alfred Wegener Institute, Bremerhaven," in *Ocean Frontiers: Explorations by Oceanographers on Five Continents*, ed. Elisabeth Mann Borgese (New York: Abrams, 1992).

26. *Tattvārtha Sūtra* 5.29–30.

27. *Tattvārtha Sūtra* 1.1.

The Nature of Nature: Jain Perspectives on the Natural World

KRISTI L. WILEY

In surveying the writings on environmental ethics published over the last quarter century, certain similarities may be seen in discussions that have arisen in the process of examining the validity of supporting an anthropocentric worldview and ideas about the nature of reality found in Jain texts written many centuries ago. The questions raised in Jainism about the natural world are not informed by the same concerns as those of twentieth-century environmentalists regarding life on this earth, which, they have observed, is being severely impacted by the ever-increasing rate of development and industrialization. Such activity pollutes the earth, water, and air to such a degree that the survival of many life-forms is in doubt. Jain *ācārya*s were concerned about the pollution of the soul by *karma,* which is understood as a type of extremely subtle matter that is attracted to and bound with the soul whenever actions are informed by passions (*kaṣāya*s). This type of pollution causes the soul to undergo transformations that give rise to *mithyātva*, or false views of reality, and causes various types of improper conduct. Engaging in conduct that minimizes volitional actions that cause harm (*hiṃsā*) and pain or suffering (*vedanā*) to other living beings also minimizes one's own suffering. Such actions result in the binding of wholesome varieties (*puṇya prakṛti*s) of karmic matter that produce feelings of bodily pleasure (*sātā-vedanīya karma*) and those *karma*s that lead to rebirth as a human (*manuṣya*) or a heavenly being (*deva*). Conversely, harmful actions cause one to bind unwholesome varieties of *karmas* (*pāpa prakṛtis*), including *karma* that produces pain (*asātā-vedanīya karma*) and those *karmas* that lead to rebirth as an animal (*tiryañca*) or hell being (*nāraki*). Understanding what in the universe is living, how living beings may be harmed, and in what manner suffering or pain is experienced by them are important in Jainism for pragmatic reasons. Therefore, it is not surprising to find detailed discussions on these subjects in ancient Jain textual sources. Here, I would like to examine certain questions that have been raised by environmentalists in light of passages from these early texts on the nature of nature, in other words, on the nature of earth, water, air, and fire, and of plants and animals.

Let us begin with the questions raised by Richard Sylvan (Routley) in his essay entitled "Is There a Need for a New, an Environmental, Ethic?" published in 1973.[1] He notes that in an attempt to move beyond the "dominant tradition" in Western ethical views regarding a person's relationship with nature, in which "nature is the dominion of man and he is free to deal with it as he pleases (since—at least on the Stoic-Augustine view—it exists only for his sake)," toward an environmental ethic,[2] a "modified dominance position" has been formulated in which "one should be able to do what he wishes providing (1) that he does not harm others and (2) that he is not likely to harm himself irreparably."[3] He observes that there are certain problems with this position, namely, "what counts as harm or interference" and who constitutes "others." As he notes, it makes a great deal of difference whether "others" is interpreted as "other humans" or "other sentient beings" and whether "future others" are included in either of these categories.

In support of his position on the rights of animals, Peter Singer provides one interpretation of "others" by quoting from the writings of Jeremy Bentham at the close of the eighteenth century.[4]

> The day *may* come when the rest of the animal creation may acquire those rights which never could have been withholden from them but by the hand of tyranny. . . . It may one day come to be recognized that the number of the legs, the villosity of the skin, or the termination of the *os sacrum*, are reasons equally insufficient for abandoning a sensitive being to the same fate. What else is it that should trace the insuperable line? Is it the faculty of reason, or perhaps the faculty of discourse? . . . But suppose they were otherwise, what would it avail? The question is not, Can they reason? nor Can they *talk*? but, *Can they suffer*?[5]

Singer concludes that

> If a being suffers, there can be no moral justification for refusing to take that suffering into consideration. No matter what the nature of the being, the principle of equality requires that its suffering be counted equally with the like suffering—in so far as rough comparisons can be made—of any other being. If a being is not capable of suffering, or of experiencing enjoyment or happiness, there is nothing to be taken into account. This is why the limit of sentience (using the term as a convenient, if not strictly accurate, shorthand for the capacity to suffer or experience enjoyment or happiness) is the only defensible boundary of concern for the interests of others. To mark this boundary by some characteristic like intelligence or rationality would be to mark it in an arbitrary way.[6]

The efforts of these environmentalists have been aimed at expanding our realm of concern from that of harm to humans (anthropocentrism) to one that includes harm to animals (biocentrism). They have justified their position on the basis that some animals have awareness or an ability to experience pleasure and pain even though they may not be capable of complex reasoning. In commenting on Singer's essay, J. Baird Callicott notes that "the minimum consideration one asks of others is not to be harmed by them," in other words, not "to be hurt, to be caused to suffer." He observes that in using sentience—"the capacity to experience pleasure and pain"—as a guideline and defining sentient beings—those who are "responsive to or conscious of sense impressions" or those who are "aware"[7]—as the "moral base class" or "criteria for moral standing," one should include all vertebrates at the very least.[8] In his opinion, this would exclude plants and "nonliving parts of ecosystems . . . such as the soil, water, and air."[9]

Moral consideration could be extended to include "entities and systems of entities heretofore unimagined (such as the biosphere itself)," if one were to adopt the view of Kenneth E. Goodpaster that "being alive" should be the basis for determining moral consideration rather than sentience or "the capacity to suffer,"[10] "since nonsentient living things may also intelligibly be said to have interests, and if so, they may be directly benefited or harmed—even though harming them may not hurt them, may not cause them consciously to suffer."[11] A separate (but related) question from establishing a "criterion of moral considerability" is the "criterion of moral significance," whether, for example, "trees deserve more or less consideration than dogs, or dogs than human persons."[12] Paul W. Taylor has expanded the definition of harm to the environment beyond the criterion of sentience by stating that "all living things are 'teleological centers of life.' An organism's *telos* (Greek for 'end, goal') is to reach a state of maturity and to reproduce. Our actions can interdict the fulfillment of an organism's *telos*, and to do just that is to harm it."[13]

Systems or holistic ecologists have proposed "ecosystem-centered ethical systems" (ecocentrism) to address ecological concerns that relate to nature when viewed as a community or ecosystem—for example, the biotic community composed of plants and animals, soils and waters. Proponents of these theories believe that biocentric criteria are inadequate for justifying that moral consideration be given to an ecosystem (provided that one accepts that such exists), in part because it includes entities that do not meet any of the above conditions since, in their view, earth, water, and air are not living. For this reason, Holmes Rolston III has proposed a system of values whereby "natural wholes, such as species and ecosystems, possess an intrinsic value derived from the baseline intrinsic value of living organisms and thus enjoy only derivative moral considerability."[14] For example, water or air would be given moral consideration because plants, animals, and humans are dependent on them for sustaining life.

From a Jain perspective, a justification for the preservation of the environment need not be based on earth, water, and air having only derivative value in their support of life. Rather, along with fire, they should be accorded moral consideration in their own right. Each of these individual elements can form the physical (*audārika*) body for a soul (*jīva*), which may be distinguished from all other existents by the quality (*guṇa*) of consciousness or awareness (*caitanya*).[15] A soul, so embodied, is a living being that is aware and that experiences pleasure and pain through its single sense of touch. Taking earth-bodied beings as an example, there are descriptions in Jain texts of the different types of earth-bodies that are formed, the minimum and maximum sizes of earth-bodies, the maximum length of time that a soul may be embodied, in birth after birth, in an earth-body before taking birth as another life-form, and the possible destinies in the life to come for a soul so embodied. In addition, there are also discussions about how one-sensed beings interact with other living beings and experience the world around them.[16]

Before beginning our investigation into the details of the lives of these beings, we should first understand the range of possible birth states according to Jain sources. There are four main destinies (*gati*s) for souls: as human beings (*manuṣya*s), heavenly beings (*deva*s), hell beings (*nāraki*s), and animals and plants (*tiryañca*). The latter, which incorporates all life-forms not included in the former three categories, is subdivided according to the number of senses or modalities of experiencing the world. In addition to those one-sensed beings mentioned above, all types of plants or vegetation (*vanaspati*) are in this category, including the *nigoda*, a minute form of vegetable life that is characterized by innumerable souls sharing a common body which, in turn, is embodied in other forms of life, including the bodies of human beings.[17] Two-sensed beings, having touch and taste, include worms, leeches, mollusks, weevils, and so on. Three-sensed beings, with the sense of touch, taste, and smell, include ants, fleas, termites, centipedes, and the like; those with four senses (additionally, sight) include wasps, flies, gnats, mosquitoes, butterflies, moths, scorpions. Five-sensed beings, those having the ability also to hear, include aquatic animals (e.g., fish, tortoise, crocodile), winged or aerial animals (birds), and terrestrials, including quadrupeds (e.g., horses, cows, bulls, elephants, lions) and reptiles (*parisarpa*).[18]

In discussing those beings whose bodies are the individual elements, we also must make a distinction between four technical terms found in Jain texts.[19] Here, we are only talking about those beings that currently have a physical body (*audārika śarīra*) that is earth (*pṛthivī-kāyika*), or water (*āpkāyika*), or fire (*tejokāyika*), or air (*vāyukāyika*). We are not talking about "earth" (*pṛthivī*), and so forth, that is not presently serving as a body for a soul and therefore is devoid of consciousness (*acetanā*). Nor are we discussing an "earth body" (*pṛthivīkāya*), that which in the recent past has served as the earth body for a soul but which has been recently abandoned by it, like the body of a person who has died. It does not include a soul currently in the process of transmigration that, upon its arrival at the locus of rebirth, will begin to grasp earth in order to form an earth-body (*pṛthivī-jīva*).[20] The latter three are excluded because the "earth" (*pṛthivī*) and an "earth body" (*pṛthivīkāya*) are nonliving material existents (*ajīva pudgala*) since they lack a soul, while a soul that is to become earth-bodied after transmigration (*pṛthivījīva*) is living, but at that moment lacks a physical body (*audārika śarīra*) of any sort.[21]

One-sensed beings should not be viewed as primitive forms of life whose souls are in the initial stages of a progressive linear evolutionary development into two-sensed life-forms, and so forth. Jains maintain that some of the infinite number of uncreated eternal souls that inhabit the occupied universe (*loka-ākāśa*) have been embodied since beginningless time as *nigodas*, the least developed of living beings. Certain of these souls have left the *nigoda* state and are currently embodied in other forms of life. And among those that have taken birth as humans, some have attained permanent release from the cycle of birth and death (*mokṣa*). However, there is no certainty that a soul will ever leave the *nigoda* state of embodiment for, unlike all other forms of life, the time that a soul may be repeatedly embodied as a *nigoda* is unending (*ananta*). Nor must the transition from a one-sensed being to other forms of life be gradual or linear. It is possible, for example, for a soul that has only been embodied as a *nigoda* to be born in its next life as a human. And according to Śvetāmbara accounts, it is possible for this soul to attain *mokṣa* in its first and only embodiment as a human.[22] Therefore, a soul that has attained *mokṣa* may never have been embodied as a two-sensed, three-sensed, four-sensed, or five-sensed animal. Conversely, humans may be born as one-sensed beings in their next birth.[23] Given the laws of *karma*, it is quite possible that a soul currently embodied as a one-sensed being has been embodied as a human some time in the past, and this soul may now be experiencing the effects of *karma* from actions undertaken as a human.

But in what sense are these one-sensed beings understood to be living and to be experiencing the effects of *karma*? Birth as a one-sensed being is attained by the fruition (*udaya*) of those *karmas* that, at the time of death of its current physical body, cause the transition of the soul to its next place of birth[24] where the soul begins to form a new physical body through the fruition of the *nāma karma* that forms a body with one sense (*ekendriya śarīra-nāma karma*). If the soul is to be earth-bodied, a specific set of subvarieties (*uttara-prakṛtis*) of *nāma karmas* comes into fruition simultaneously and causes the formation of a separate body (*prateyka śarīra nāma karma*) by attracting particles of earth, transforming them, and binding them together to form a body of a specific size and shape.[25] Until the time of death, certain of these *nāma karmas* will continue to rise, causing the continuous influx of matter to maintain the body.[26]

Like other beings who have not attained omniscience (*kevalajñāna*), once in each life all one-sensed beings must bind *āyu* (longevity) *karma*, which establishes the maximum length of life and determines whether one's next birth will be as a human, animal, heavenly being, or hell being. In this regard, differences among the various types of one-sensed beings have been noted. The maximum length of life for an earth-bodied being is different, for example, from that of a water-bodied being.[27] And there are differences in the subvarieties of *āyu karma* that the various categories of one-sensed beings may bind. Although it is possible for the soul of a plant, earth-bodied being, or water-bodied being to bind human (*manuṣya*) *āyu* and thus be reborn as a human in its next life, a fire-bodied being or air-bodied being can only bind the *āyu karma* that causes rebirth as an animal or a plant (*tiryañca āyu*).[28] I have found no explanation in the commentaries for such distinctions; however, these notions could possibly be related to the greater amount of *hiṃsā* that air- and fire-bodied beings are capable of causing, especially in cooperation with each other.

Along with various forms of vegetable life, earth-bodied, water-bodied, fire-bodied, and air-bodied beings all develop four life-forces, or vitalities (*prāṇas*), from the rise of *āyu* and *nāma karmas*: the vitality of the strength or energy of the body (*kāyabala prāṇa*); the vitality of respiration (*ucchvāsaniśvāsa prāṇa*); the vitality of life span (*āyuḥ prāṇa*); and the vitality of the sense of touch (*sparśanendriya prāṇa*). However, they are unable to develop other vitalities, such as the sense of taste and the ability of speech (or the ability to make sounds) of two-sensed beings, or the additional vitality of the sense of smell of three-sensed beings, or the sense of sight of four-sensed beings, or the sense of hearing of five-sensed beings, or the sense of rationality of five-sensed rational beings (*pañcendriya saṃjñīs*).[29] In the case of one-sensed beings, these four vitalities of energy, respiration, life span, and touch cannot be detected by a person through sense perception. Therefore, one of the chief mendicant disciples (*gaṇadharas*) of Mahāvīra, the twenty-fourth Tīrthaṅkara, asks him: "We know and observe the inhalation and exhalation, the breathing in and breathing out, of those living beings who are two-sensed, three-sensed, four-sensed, and five-sensed, but we do not know or observe this in the case of one-sensed beings, from earth-bodied beings through *vanaspati*. Do beings that are one-sensed also inhale and exhale, breathe in and breathe out?" Mahāvīra replies, "Oh Gautama, these living beings with one sense also inhale and exhale, breathe in and breathe out."[30]

Jain texts mention four instincts (*saṃjñās*) that are present even in one-sensed beings. Craving for food (*āhāra-saṃjñā*) is the most primary of these instincts. Other instincts include fear (*bhaya-saṃjñā*), the desire for reproduction (*maithuna-saṃjñā*), and the desire to accumulate things for future use (*parigraha-saṃjñā*).[31] Gautama inquires of Mahāvīra, "Do earth-bodied beings desire nourishment (*āhāra*)?" Mahāvīra says, "Yes, they desire nourishment. At all times and without interruption the desire for food arises in them. It is transformed repeatedly in various ways by the organ of touch in the form of pleasant and unpleasant feelings."[32] In the case of a one-sensed being, which lacks a mouth, nourishment consists of matter that is assimilated through the surface of the entire body. Such intake is considered involuntary, in contrast with the voluntary consumption of "food by morsel" by two-sensed beings with a mouth, which accept or reject food based on the sense of taste.

It is clear from other passages in these texts that one-sensed beings interact with the world around them. With respect to such activities, Gautama asks, "Do all earth bodies have similar activities (*samakiriyā* = Sanskrit, *samakriyā*)?" Mahāvīra answers, "Yes, they all have similar activities." "Why so?" "All earth bodies are with deceit (*māyā*) and wrong outlook (*mithyātva*). So they have five activities, which are those arising out of endeavour (*ārambhikā kriyā*), etc., till those arising out of perverted faith (*mithyādarśaṇapratyayā kriyā*) through the sense of touch."[33] It is said that merely by breathing, earth-bodied, water-bodied, air-bodied, and fire-bodied beings, as well as plants, commit three, four, or five types of actions, while an air-bodied being, stirring part of a tree or causing it to fall down, also commits three, four, or all five actions.[34] Since they are subject to the various passions (*kaṣāyas*) of anger (*krodha*), pride (*māna*), deceit (*māyā*), and greed (*lobha*) produced by the rise of *mohanīya karma*,[35] their actions are volitional. Therefore, like those humans that have not attained omniscience and perfect conduct and are thus still subject to these four passions, activities of all one-sensed beings cause the influx and binding of new karmic matter that may be experienced in its current life and in lives to come.

And it is clearly stated that one-sensed beings experience suffering through the sense of touch. Gautama inquires of Mahāvīra, "Do all earth-bodied beings have an equal feeling of suffering (*samaveyaṇā* = Sanskrit, *samavedanā*)?" Answer: "Yes, they have an equal feeling of suffering." Why? "All earth-bodied beings are devoid of a conscious mind (*asaṃjñī*) and so they experience pleasure and pain (*vedanā*) in an indeterminate way or with the absence of positive knowledge (*aṇidāe*)."[36] In a note on this verse, K. C. Lalwani states:

> The indeterminateness of pain is signified by the word *aṇidāe*. This is so because of wrong outlook and absence of reasoning, for which, like one under the spell of a drug or drink, they do not know what they are suffering from, and how much is their suffering. They accept their suffering as *fait accompli* and are used to it. The same applies to the other one-sensed beings.[37]

In the *Bhagavatī Sūtra* it is said that an earth-bodied being experiences pain (*vedaṇā*) "as great as that of an old decrepit man whom a young strong man gives a blow on the head."[38]

Thus, according to Jain sources, whenever matter in the form of earth, water, air, or fire is embodying a soul, it constitutes a living being, which breathes, nourishes its body, and sustains life in its body. Like other beings, a one-sensed being performs actions and will experience the karmic effects of these actions. And it feels pleasure and pain through the sense of touch. As mentioned in the opening lectures of the *Ācārāṅga Sūtra*, even though hurting one-sensed earth beings may not be readily apparent through observation, a person can hurt them and cause them to suffer by cutting, striking, or killing them.[39]

Returning to the discussions of twentieth-century environmentalists, if one were to use the material in Jain texts to interpret the statement that "one should be able to do what he wishes providing (1) that he does not harm others and (2) that he is not likely to harm himself irreparably,"[40] and define "other" or "moral base class" (following Callicott's observations) as those with sentience ("the capacity to experience pleasure and pain") or those who are "aware,"[41] one would include not only nonrational five-sensed animals (*pañcendriya asaṃjñī*s), but also life-forms with just one sense. In discussing the concept of moral rights or those beings that deserve moral consideration, Joel Feinberg has stated that "a being without interests is a being that is incapable of being harmed or benefited, having no good or 'sake' of its own," and that "interests" logically supposes desires or "wants" or "aims."[42] It is clear from the above passages that in the Jain worldview one-sensed beings can be harmed. They have the capacity to experience pleasure and pain, and they are aware because they have a *jīva*, or soul, whose defining characteristic is awareness. They also have "desires" because they experience the effects of *mohanīya karma*s, which generate passions (*kaṣāya*s) of attraction (*rāga*) and aversion (*dveṣa*). Because of other *karma*s, they are subject to the instincts (*saṃjñā*s) of fear (*bhaya-saṃjñā*) and the desire for food (*āhāra-saṃjñā*), for reproduction (*maithuna-saṃjñā*), and for the accumulation of things for future use (*parigraha-saṃjñā*).[43] Kenneth Goodpaster has expressed similar ideas regarding plants:

> There is no absurdity in imagining the representation of the needs of a tree for sun and water in the face of a proposal to cut it down or pave its immediate radius for a parking lot. . . . In the face of their obvious tendencies to maintain and heal themselves, it is very difficult to reject the idea of interests on the part of trees (and plants generally) in remaining alive.[44]

In the context of Jain sources, this would include earth-bodied, water-bodied, fire-bodied, and air-bodied beings because they also need to nourish their bodies in order to stay alive.

There is also some similarity between Taylor's definition of living beings as "teleological centers of life," whose goal "is to reach a state of maturity and to reproduce,"[45] and harm as interference with the fulfillment of an organism's *telos* and the Jain definition of *hiṃsā* as harm to the life forces, or *prāṇa*s, including the life force of longevity (*āyu*).[46] In contrast with a speciesist view where "the interests of others matter only if they happen to be members of his own species," Jain *ācārya*s have maintained that "All beings are fond of life, like pleasure, hate pain, shun destruction, like life, long to live. To all life is dear."[47] "All breathing, existing, living, sentient creatures should not be slain, nor treated with violence, nor abused, nor tormented, nor driven away."[48]

In examining the validity of an anthropocentric worldview, the question has been raised by Paul Taylor: "In what sense are humans alleged to be superior to other animals?"[49] According to the teachings of Jainism, humans are different from all other beings because they have a capacity that all others lack: the ability to attain omniscience (*kevalajñāna*) and permanent release from the beginningless cycle of death and rebirth (*mokṣa*).[50] Jains believe that five-sensed rational animals can attain true spiritual insight (*samyak-darśana*), the first step toward *mokṣa*. It is said that animals who have attained this insight can observe restraint with respect to killing, and so forth, and even refuse food at the approach of death. Thus, they are able to follow a mode of conduct equivalent to that of a person who has accepted the lay vows (*aṇuvrata*s). However, an animal is incapable of attaining more advanced states of spiritual purity that are a prerequisite for *mokṣa*.[51] Only a human being has the ability to undertake the physical and mental austerities necessary for removing all *ghātiyā* (destructive) *karma*s, which prevent a person from realizing omniscience (*kevalajñāna*) and perfect conduct, from experiencing the true nature of the soul, and from attaining release from the cycle of death and rebirth.

In Jainism, the spiritual well-being of a person is tied to the physical well-being of all forms of life. This is reflected in the vows of restraint (*vrata*s) that a Jain may formally take to refrain from harmful actions (*ahiṃsā*), from telling lies (*satya*), from stealing (*asteya*), from inappropriate sexual activity (*brahmacarya*), and from possessiveness (*aparigraha*). By observing these vows, a person tries to refrain from actions that cause harm to other beings. For a person who has accepted the lay vows (*aṇuvrata*s), this means not harming beings with two, three, four, and five senses. This consideration for the welfare of animals among Jains is demonstrated by their emphasis on vegetarianism, their preference for those occupations that minimize harm to living beings, and by the establishment of special refuges for animals, called *pinjrapoles*. Although the efforts of these institutions are focused primarily on protecting domestic herd animals, such as goats, sheep, and cattle, other sick or injured animals may be brought to these refuges for shelter and medical treatment. Animals that are of no economic importance also are cared for here. There is often a sanctuary for birds, where food and water is provided out of the reach of predators.[52] Practices at the *pinjrapole*s, where a being is allowed to live out the life span with which it was born and to die a natural death, are in accordance with a definition of *ahiṃsā* that includes noninterference with a being's life force (*āyu prāṇa*).[53] The emphasis here is on providing an environment of protection for the preservation of life, rather than ending the pain and suffering of injured or sick animals through premature termination of life by euthanasia.

The acceptance of the mendicant vows (*mahāvratas*), which are indicative of even higher states of spiritual purity, entails more restrictions on one's actions, because the vow of *ahiṃsā* encompasses one-sensed beings as well. Through circumscribed actions mendicants avoid harming plant life by not walking on greenery or touching a living plant; air-bodied beings by not fanning themselves; fire-bodied beings by not kindling or extinguishing fire; water-bodied beings by not swimming, wading, using water for bathing, or drinking water that has not been properly boiled; and earth-bodied beings by not digging in the earth.[54] As noted by Padmanabh S. Jaini:

> Perhaps every culture teaches its children to behave with regard for the well-being of other persons and of domestic animals. The normal socialization process, however, provides little or no basis for extending this consideration to the single-sensed creatures. Hence the Jaina mendicant must put forth a tremendous effort of mindfulness, consciously establishing a totally new pattern of behavior for which his prior training has in no way prepared him. Undertaking *ahiṃsā* and the other great vows forces him to become constantly aware of his every action, always on guard against the possibility of committing an infraction.[55]

This interconnection between spiritual well-being and physical well-being is demonstrated in the practice of asking forgiveness for past transgressions (*ālocanā*) from all living beings.

> I want to make *pratikramaṇa* for injury on the path of my movement, in coming and in going, in treading on living things, in treading on seeds, in treading on green plants, in treading on dew, on beetles, on mould, on moist earth, and on cobwebs; whatever living organisms with one or two or three or four or five senses have been injured by me or knocked over or crushed or squashed or touched or mangled or hurt or affrighted or removed from one place to another or deprived of life—may all that evil have been done in vain [*micchāmi dukkaḍaṃ*].[56]

> I ask pardon of all living creatures, may all of them pardon me, may I have friendship with all beings and enmity with none.[57]

Such concerns for the well-being of even the most minute life-forms, accompanied by voluntary restraints on the accumulation of possessions and limiting the consumption of finite natural resources, accords well with a responsible environmental ethic. In the words of Harold Coward:

> To harm any aspect of nature—be it air, water, plants, or animals—is tantamount to harming oneself. Thus there is a clear and unambiguous environmental ethic within Indian thought. The fact that such an ethic has not protected South Asia from the environmental problems of modern industry and agriculture suggests that it has not been sufficiently understood and applied.[58]

However, limitations are encountered in using Jain sources to examine some of the more difficult problems in environmental ethics. One is how to establish a "criterion of moral significance," which

> aims at governing *comparative* judgments of moral "weight" in cases of conflict. Whether a tree, say, deserves any moral consideration [i.e., the criterion of moral considerability] is a question that must be kept separate from the question of whether trees deserve more or less consideration than dogs, or dogs than human persons. We should not expect that the criterion for having "moral standing" at all will be the same as the criterion for adjudicating competing claims to priority among beings that merit that standing.[59]

Could a measure such as "like suffering" be used as a standard? In justifying his position on sentience as a guideline for determining the interests of others, Singer has stated, "No matter what the nature of the being, the principle of equality requires that its suffering be counted equally with the like suffering—in so far as comparisons can be made—of any other being."[60] According to Jain sources, is there a difference between suffering experienced in an "indeterminate" manner and that experienced by beings with the mental capacity to reason, to reflect on the past, and think about the future? Instructive in this regard is a passage about hell beings in the *Bhagavatī Sūtra* in which Gautama asks Mahāvīra, "Do all infernal beings suffer an equal pain?" Answer: "This is not necessarily so." Why? "The infernal beings are of two types. They are those with consciousness (*saṃjñī*) and those without consciousness (*asaṃjñī*). Those who have consciousness have great pain, and those who are without consciousness have little pain."[61] According to this statement, beings with "like suffering" would be divided into two groups: 1) five-sensed rational beings, including both animals and humans; and 2) one-sensed to five-sensed nonrational beings. Therefore, if one were to use this division as a guideline for a priority of moral significance, then humans and animals would rank first, with other beings ranked equally below this. However, the way in which pain is understood to be experienced by rational and nonrational beings apparently was not the criterion used for defining conduct appropriate for householders or mendicants. Rather, the dividing line between refraining from harming two-sensed beings on the part of laypeople and one-sensed beings on the part of mendicants probably reflects practical limitations on the degree to which such restraints could be practiced on a daily basis by members of these two communities.

Jains have considered a similar question in trying to define the amount of *hiṃsā* that one accrues from harm done to other living beings. In discussing why it is important for Jains to understand the number of vitalities (*prāṇa*s) that different types of living beings have, J. L. Jaini states, "the degree of sin would depend upon the number of vitalities and their comparative strength, to which injury is caused. The knowledge of the varying number of vitalities possessed by souls in their various conditions of life enables one to judge the extent of injury he is likely to cause in his actions."[62] By this definition, there would be a hierarchy of *hiṃsā*, or a sliding scale of spiritual harm, with progressively less harm to one's soul from causing injury to a five-sensed rational animal, a five-sensed nonrational animal, and so forth. The least problematic would be injury to a one-sensed being. While this idea does not directly translate into "more or less significance," the karmic consequences from harming a five-sensed rational being and a one-sensed being are not considered to be equal.

It is unclear to me how ideas expressed in Jain texts might be used in support of holistic views of environmental ethics. Although ideas expressed in these texts justify an extension of the "circle of moral considerability," this is done from an individualistic perspective. These sources focus "concern on particular items, whether they be persons, animals, living things, or natural items." However, if one

> introduces a holistic element . . [then] whole ecosystems, the biosphere, and even the universe as a whole are morally considerable and the particular individuals which constitute them are themselves only insignificantly, if at all, considerable. . . . These large systems exhibit

> sufficient organization and integration to count as alive, as having a good of their own or, less controversially, as possessing intrinsic value.[63]

I can see no evidence in Jain texts for the devaluation of individuals within a given class, be it humans or one-sensed beings, in favor of the group or species, especially considering the Jain conception of the soul. Although souls share certain common characteristics, such as consciousness, the soul of each being is a separate entity, with its own unique accumulation of karmic matter. It retains its own identity and isolation even in *mokṣa* and does return to or become part of a cosmic soul. Nor can I see strong evidence in support of organization or integration of larger systems, such as ecosystems or the universe or viewing these entities as living, unless one considers the shape of a "cosmic man," which is sometimes poetically used to depict the boundaries of the occupied universe (*loka-ākāśa*). However, a stronger case could be made for a part/whole human/universe correspondence based on material found in early brahmanical texts, such as the Puruṣa-sūkta hymn.

I am also at a loss as to how to explain environmental harm in the context of still another category of beings: subtle (*sūkṣma*) one-sensed beings. All of the discussions up to now have been about beings with bodies (*śarīra*s) that are "gross" (*bādara* or *sthūla*). Such bodies are called *ghāta śarīra*s because they are composed of matter that can be obstructed by or harmed by other matter and that can itself obstruct or harm other objects. All beings with two or more senses have bodies composed of gross matter. However, in the case of one-sensed beings, depending on the specific subvariety of *nāma karma* that comes into fruition at the time of its "birth" when the body begins to form, the external body (*audārika śarīra*) may be composed of either gross matter or "subtle" or "fine" (*sūkṣma*) matter. A body composed of fine matter is nonobstructive (*aghāta*) because it neither obstructs nor is obstructed by other objects. According to the *Gommaṭasāra*: "A fine body can pass through any kind of matter. . . . They are indestructible or non-obstructive because nothing can kill them and they can kill nothing. They die a natural death at the exhaustion of their age (*āyu*) karma."[64] "Gross bodies need support but fine bodies need no support and exist everywhere (in the occupied universe) with nothing intervening between them."[65] Although this category of beings is mentioned in the Śvetāmbara and Digambara texts that discuss the soul and *karma*, to the best of my knowledge, there is no separate discussion of them in the texts devoted to the conduct of mendicants. They must experience pain because they are subject to the rise of *asātā vedanīya karma*. But apparently the premature rise of this *karma* (*udīraṇā*) cannot be caused by external factors, as is the case with gross-bodied beings. If one understands that these beings cannot be harmed by others, then to what extent can one equate pollution of the environment with harm to one-sensed beings in the form of earth, water, fire, and air? It would seem that even within Jainism there are limits to harm. But it would be safe to say that except for these very subtle forms of life, the environment in the form of one-sensed beings can be harmed by the actions of humans. What we understand as "earth" is that which can be detected via the senses of touch, smell, taste, and sight. Such matter is classified as gross matter in Jainism, and whenever this matter is currently embodying a soul, it constitutes a gross one-sensed being that can be harmed by the actions of others.

Texts that discuss the nature of reality and what is appropriate conduct for mendicants provide strong evidence for expanding our circle of moral consideration to include earth, water, and air in their own right. However, one should not expect these same texts to provide guidance in deciding what should be done about environmental harm caused in the course of living a householder's life.[66] Instead, one should look to stories in Jain narrative literature, such as the *Ādipurāṇa* of Jinasena.[67] Here, it is said that long ago in Bharatakṣetra, living conditions were such that people were supplied with food and other necessities of life by wish-fulfilling trees (*kalpavṛkṣas*).[68] From an environmental perspective, this was an ideal state of affairs because there was no agriculture to affect the natural wild species of plants and there was no damage to the earth, water, and air from the manufacturing of goods for human consumption. However, unlike certain other locations in the universe that are not subject to cyclical time, there came a point in the descending cycle of time (*avasarpiṇī*) when this was no longer the case. During the time of Ṛṣabha, the first Tīrthaṅkara of our *avasarpiṇī*, conditions were such that food and the other necessities of life were no longer plentiful, and social problems arose as people competed with each other for dwindling resources. As king, Ṛṣabha could have done nothing, since it was inevitable that things would only get worse over time as the descending cycle continued. However, this is not the course of action that he took. Instead, he taught the people agriculture and crafts so they could provide themselves with the things that previously had been acquired with no effort, even though all of these activities, by their very nature, must have been harmful to one-sensed beings. He achieved social stability by establishing the occupational social divisions (*varṇa*s) of Kṣatriyas, Vaiśyas, and Śūdras. It was one of his sons, King Bharata, who established the Brahmin *varṇa* after Ṛṣabha had renounced the world. Bharata decided which individuals should be included in this new *varṇa* on the basis of conduct. His selection was made by observing those who, when faced with a choice of the direction in which to walk toward him, chose the path of lesser harm by not trampling on the grass.

Throughout history Jains have been faced with making choices in their daily lives as they decide the extent to which they should follow the guidelines laid out for them by Jain *ācārya*s in texts detailing conduct appropriate for the lay community.[69] In commenting on the list of fifteen trades forbidden to Jains as outlined in the Śvetāmbara *śrāvakācāra* texts, R. Williams has said:

> The eternal dilemma of Jainism in laying down an ethos for the layman has been well put by Āśādhara. The lay estate . . . cannot exist without activity and there can be no activity without the taking of life; in its grosser form this is to be avoided sedulously but the implicit part of it is hard to avoid. . . . [A]t least the keeping of animals and contact with any destructive implements are to be eschewed.[70]

However, in these same texts, one is reminded of the compromises of a household life, which is equated with life in a slaughterhouse (*sūna*). Pounding, grinding, cooking, cleaning, and sweeping all impede the path to *mokṣa* because they cause the destruction of living beings.[71] Over the centuries, Jains have decided whether to formally accept any or all of the *aṇuvrata*s and to what degree they would curtail their activities. What limits on the acquisition of property and possessions might a person voluntarily abide by? To what extent might a person accept a vow to limit travel (*dig-vrata*) for a specified period of time since,

> Like a heated iron sphere the layman will inevitably, as a result of *pramāda*, bring about the destruction of living creatures everywhere, whether he is walking, or eating, or sleeping, or working. The more his

> movements are restricted, the fewer *trasa-jīvas* [those that are capable of moving from one place to another, beginning with two-sensed beings] and *sthāvara-jīvas* [those that cannot move on their own, all one-sensed beings] will perish.[72]

In this context, the question is what trade-offs would a person make between the purification of the soul, on the one hand, and unrestricted activity on the other?

Today, there are practical considerations that need to be addressed when one contemplates putting into practice an environmental ethic that accords moral standing to what in Jainism are one-sensed beings. If one accepts earth, water, fire, and air in this category, then some way needs to be found to balance their well-being with that of humans. For those who have not renounced the household life, a definition of well-being would, in general, include a certain degree of physical comfort that is afforded by having access to electricity, running water, mechanized transportation, adequate health care, shelter, and clothing. These comforts are not possible without development and industrialization, which causes harm to the earth, water, and air. As Harold Coward has observed, "Seeing earth, air, and water as beings in different forms, as Jaina *karma* theory does, provides an ethic that rejects the ruthless exploitation of natural resources that modern industrial development practices and the environmental pollution (including disasters like Bhopal) that result." However, "the Jaina conception of *karma* theory may be too radical, in spite of its logical consistency, to be taken seriously by modern India."[73]

Nonetheless, Jains who accept as authoritative the nature of reality as described in their ancient texts, who are still conforming to the standards of conduct laid down in centuries past—be it vegetarianism or not eating after dark—must not ignore the reality of the harm that is being done today to the earth, water, and air in the cities in which they live from mechanized transportation and forms of production that scarcely could have been imagined by the *ācārya*s in centuries past. As passages in Jain narrative literature and *śrāvakācāra* texts illustrate, when faced with alternatives, avoiding harm to one-sensed beings whenever possible, to whatever degree possible, is still the ideal to strive for, even on the part of householders. A Jain, therefore, would not support the following statement: "while technically they [plants and other barely living beings] may be morally considerable, practically they may fall well below the human 'threshold' of moral sensitivity. Thus we may forever be unable actually to take the interests of all the living things that our actions affect into account as we make our day-to-day practical decisions."[74] As long as there are Jain monks and nuns in sufficient numbers as there are in India today, who have taken vows not to harm one-sensed beings, Jain laypeople, who have traditionally supported them with great devotion, will undoubtedly remain aware of one-sensed beings and will conduct themselves in such a way as will be acceptable to the mendicant community. In so doing, they will try to minimize the use of and violence toward one-sensed beings.

In earlier times, the question has been raised regarding what could be done to offset the spiritual harm caused to one-sensed beings by the activities involved in leading a household life. According to Āśādhara, impediments to spiritual well-being caused by harm done to one-sensed beings could be eliminated by almsgiving to ascetics.[75] In modern times, when it is not practical to avoid harm to the earth, water, and air, what actions could be undertaken to compensate for harm done to the environment from the way in which we live today? This question should be pondered with a view toward Jain perspectives on the nature of nature and the one-sensed beings that constitute the environment of the earth on which we live.

Notes

1. Richard Sylvan (Routley), "Is There a Need for a New, an Environmental, Ethic?" in *Environmental Philosophy: From Animal Rights to Radical Ecology*, ed. Michael E. Zimmerman et al., 2d ed. (Upper Saddle River, N.J.: Prentice Hall, 1998), 17–25. Originally published in *Proceedings of the Fifteenth World Congress of Philosophy*, no. 1 (Varna, Bulgaria, 1973), 205–10.

2. Ibid., 18.

3. Ibid., 20, quoting from Paul W. Barkley and David W. Seckler, *Economic Growth and Environmental Decay: The Solution Becomes the Problem* (New York: Harcourt Brace Jovanovich, 1972), 58.

4. Peter Singer, "All Animals Are Equal," in *Environmental Philosophy*, ed. Zimmerman et al., 26–40. Originally published in *Philosophic Exchange* 1, no. 5 (summer 1974): 243–57.

5. Singer, quoting Jeremy Bentham, *Introduction to the Principles of Morals and Legislation*, chap. 17, in "All Animals Are Equal," 30.

6. Singer, "All Animals Are Equal," 31.

7. As defined in *Merriam-Webster's Collegiate Dictionary*, 10th ed., s.v.

8. J. Baird Callicott, "Introduction to Part One: Environmental Ethics," in *Environmental Philosophy*, ed. Zimmerman et al., 10.

9. Ibid., 11.

10. Kenneth E. Goodpaster, "On Being Morally Considerable," in *Environmental Philosophy*, ed. Zimmerman et al., 58; originally published in *The Journal of Philosophy* 75, no. 6 (June 1978): 308–25. "As far as I can see, X's being a living thing is both necessary and sufficient for moral considerability" (p. 60). In this article, he leaves questions regarding the criterion of moral significance unanswered. See note 59 below.

11. Callicott, "Introduction to Part One: Environmental Ethics," 11.

12. Goodpaster "On Being Morally Considerable," 59.

13. Paul W. Taylor, quoted and commented upon in Callicott, "Introduction to Part One: Environmental Ethics," 11.

14. Ibid., 15.

15. For the other existents in Jainism, see Padmanabh S. Jaini, *The Jaina Path of Purification* (Berkeley and Los Angeles: University of California Press, 1979), 90–91 and 97–106.

16. These details are found in a number of sources, including Śvetāmbara canonical texts, that discuss the nature of the soul and the parameters for its embodiment and rebirth, e.g., the *Bhagavatī Sūtra*, *Prajñāpanā Sūtra*, and *Jīvājīvābhigamana Sūtra*; in texts devoted to *karma* theory, e.g., *Karmaprakṛti* and the *Karmagranthas*; and in Digambara texts on these subjects, e.g., *Gommaṭasāra*, *Ṣaṭkhaṇḍāgama* and its commentary entitled *Dhavalā*.

17. The body of a *nigoda* is formed by the operation of the common body-making *karma* (*sādhāraṇa śarīra nāma karma*), and such bodies are called "group souled" (*sāmaṇya*). A *nigoda* is defined as "that which is always the abode of infinite souls. All souls occupying the same body share common food and common respiration. Souls in such embodiments have extremely short life spans. However, once formed, a

nigoda body continues to exist for very long periods of time. Within this body, at every instant, infinite souls are said to die and new souls take birth." See *Gommaṭasāra Jīvakāṇḍa,* 2 vols., Jñānapīṭha Mūrtidevī Jaina Granthamālā, Prakrit Grantha, nos. 14–15 (New Delhi: Bhāratīya Jñānapīṭha, 1978–1979) [= *GJK*], 191–93; and comments by J. L. Jaini, trans., *Gommaṭasāra Jīvakāṇḍa,* Sacred Books of the Jainas, vol. 5 (Lucknow: The Central Jaina Publishing House, 1927; reprint, New Delhi: Today and Tomorrow's Printers and Publishers, 1990), 118–19.

18. *Tattvārtha Sūtra* 2.22–2.23.

19. See *Sarvārthasiddhi* 2.13 and the commentary on *GJK* 182.

20. Because the soul is not considered omnipresent (*vibhū*), it must move from the locus where the death of the physical body occurs to the place of rebirth. Two technical terms are used for this transitional state: *apāntarāla-gati* (see Helmuth von Glasenapp, *The Doctrine of Karma in Jain Philosophy,* translated from the German by G. Barry Gifford [Bombay: Bai Vijibhai Jivanlal Pannalal Charity Fund, 1942], 33), and *vigraha-gati.* In contrast with those religious traditions that understand a lengthy period of transition, in Jainism it is described as virtually instantaneous, lasting at the most only four moments.

21. Although at death the soul leaves the physical (*audārika*) body, during transmigration it still retains the dimensions of that body due to the operation of *ānupūrvī nāma karma.* Prior to *mokṣa,* a soul is always embodied because the soul transmigrates with two bodies: the *karmaṇa śarīra,* which is composed of all accumulated karmic matter, and the *taijasa śarīra,* which provides heat for the other bodies.

22. According to Śvetāmbara sources, this was the destiny of the soul of Marudevī, the mother of the first Tīrthaṅkara of our current descending cycle of time (*avasarpiṇī*) For details, see Padmanabh S. Jaini, "From Nigoda to Mokṣa: The Story of Marudevī," in *Proceedings from the International Conference on Jainism and Early Buddhism, Lund University, June 4–7, 1998* (forthcoming). *Nitya nigoda* or *avyavahārika nigoda* are the designations for a soul that has never taken birth in another state of existence.

23. According to Digambara sources, this was the destiny of the soul of the Ājīvika Makkhali Gosāla and of several others who preached false doctrines, who were reborn as *nigoda*s in their next life. *Caturgati nigoda, itara nigoda,* or *vyavahārika nigoda* are the designations for a soul that in the past has been embodied in other life-forms.

24. These include animal life span *karma* (*tiryañca-āyu karma*) and the associated *nāma karma*s that bring about such a birth, including *tiryañca gati-nāma karma* and *tiryañca ānupūrvī-nāma karma.*

25. See *Bhagavatī Sūtra* [= *BhSū*] *śataka* 19, *uddeśaka* 3, pp. 834–40 (762a–765b), and summary by Jozef Deleu, *Viyāhapannatti (Bhagavaī): The Fifth Anga of the Jaina Canon* (Brugge: De Tempel, Tempelhof, 1970), 248–50. *BhSū* references are numbered according to the Jain Āgama Series edition, followed by the page number in that edition: *Vyākhyāprajñapti Sūtra,* 3 vols., Jaina Āgama Series, no. 4, pts. 1–3, ed. Pt. Bechardas J. Doshi (Bombay: Shrī Mahāvīra Jaina Vidyālaya, 1978–1982). Where applicable I have given in parenthesis the page number of the Āgamodaya-Samiti edition with Abhayadevasūri's commentary as cited in Deleu, *Viyāhapannatti*

(Bhagavaī): *Vyākhyāprajñapti Sūtra*, 4 vols., Jināgama Granthamālā, nos. 14, 18, 22, 25 (Beawar, Rajasthan: Sri Agam Prakashan Samiti, 1982–1986).

According to the *Uttarādhyayana Sūtra* (36.69–126), earth-bodies may be either smooth or rough. Smooth ones are said to be of seven kinds: black, blue, red, yellow, white, pale dust, and clay, while rough ones are of thirty-six kinds: earth, gravel, sand, stones, rocks, rock-salt, iron, copper, tin, lead, silver, gold, diamond (and so forth). Their varieties, caused by (difference of) color, smell, taste, touch, figure, and place, are counted by the thousands. Water-bodied beings are of five types: pure water, dew, exudations, fog, and ice. Fire-bodied beings are of many kinds, including coal, burning chaff, fire, and flame of fire, meteors, lightning, and many other kinds. Air-bodied beings are of five kinds: squalls or intermittent winds (*utkalikā*), whirlwinds (*maṇḍalikā* = *vātolī*), thick winds (winds that blow on the oceans below the Ratnaprabhā hell or which support the heavenly Vimānas), high winds, and low winds. There is also a wind called Saṃvartaka, which, according to Devendrasūri, carries grass, etc., from outside into a particular place. Hermann Jacobi understands this wind as "the hurricane which causes the periodical destruction of the world" and notes that in the *sūtra*, more winds are implied by the word "etc." (*ādi*). Translation follows Hermann Jacobi, trans., *Jaina Sūtras*, pt. 2 (Oxford: Oxford University Press, 1895; reprint, Delhi: Motilal Banarsidass, 1989), pp. 213–18, and notes 3–5, p. 218.

26. For the set of *nāma karma*s that operate to form the bodies of various types of one-sensed beings, see Glasenapp, *The Doctrine of Karma in Jain Philosophy*, 31–32, list numbers 23, 25a, and 26.

27. The maximum length of life, which is determined by the *sthiti* of *āyu karma*, is 22,000 years for earth-bodied beings, 7,000 years for water-bodied beings, 3 days for fire-bodied beings, 3,000 years for air-bodied beings, and 10,000 years for plants (*Uttarādhyayana Sūtra* 36.81, 36.89, 36.112, 36.123, and 36.103, respectively, as translated in Jacobi, *Jaina Sūtras*, 213–18).

28. This distinction is also reflected in possible rebirths for certain heavenly beings, namely the Bhavanapati, Vyantara, Jyotiṣka, Saudharma, and Īśāna *deva*s, who may be born as plants or earth-bodied and water-bodied beings, but not as fire-bodied or air-bodied beings. For details, see Glasenapp, *The Doctrine of Karma in Jain Philosophy*, 53.

29. See *Sarvārthasiddhi* 2.13–14 and *GJK*133. The number of vitalities ranges from four (one-sensed beings) to ten (five-sensed rational beings). For purposes of this discussion, I have excluded another category of living beings discussed in Jain texts, undevelopable beings (*aparyāpta*s), who have an extremely short life span (one-eighteenth of a pulse beat) and are unable to develop the vitalities necessary to maintain life as a one-sensed being, and so forth.

30. *BhSū*, *śataka* 2, *uddeśaka* 1, *sūtra*s 3–7, pp. 73–75 (110a). Gautama also poses a question about respiration in the case of air-bodied beings. "Do air-bodies grasp and release through breathing in and breathing out other air-bodies?" Mahāvīra replies, "Yes, air-bodies grasp and release through breathing in and breathing out other air-bodies." This passage has been interpreted to mean that the breathing in and breathing out of air-bodied beings is not in the form of bodies of air-bodied beings that are sentient (*caitanya*). This air is lifeless. See K. C. Lalwani, trans., *Bhagavatī Sūtra*, 3 vols. (Calcutta: Jain Bhawan, 1973–1980), 2:339: "Air-bodies inhale and exhale air,

but not air-bodies. In other words, the air they breathe in and out is not living air. It is without life." Lalwani (*Bhagavatī Sūtra*, 1:271) notes why this is a relevant question. "Other beings inhale and exhale what is air. But when air-bodies do the same, do they inhale other air-bodies? And if one air-body inhales another air-body, and a second air-body inhales a third, and so on, then, who inhales and who is inhaled? This leads to a fallacy. The solution is that what the air-bodies inhale and exhale are not air-bodies but non-live air. Since the air inhaled and exhaled by the air-bodies is without life, the former stands in no need to inhale and exhale."

31. *GJK* 134-139. See also *Prajñāpanā Sūtra*, chapter 8.

32. *BhSū śataka* 1, *uddeśaka* 1, *sūtra* 6.12.3–5, p. 8.

33. *BhSū śataka* 1, *uddeśaka* 2, *sūtra* 7.3, p. 19 (39a) as translated by Lalwani, *Bhagavatī Sūtra*, 1:40. The five types of activities, which are abbreviated in this *sūtra*, are enumerated in an earlier passage about hell beings (*BhSū śataka* 1, *uddeśaka* 2, *sūtra* 5.6, p. 18). They are activities that arise as a result of endeavor (*ārambhikā kriyā*), possessiveness (*pārigrahikā kriyā*), deceit (*māyāpratyayā kriyā*), non-abstinence (*apratyākhyānā kriyā*), and wrong outlook (*mithyādarśaṇapratyayā kriyā*). Deceit (*māyā*) implies the presence of the other three passions (*kaṣāya*) mentioned in the following sentence.

34. *BhSū śataka* 9, *uddeśaka* 34, *sūtras* 9–25, pp. 483–84 (491b), as translated by Deleu, *Viyāhapannatti (Bhagavaī)*, 165–66. Here, the five actions refer to those caused by means of one's body (*kāyikī*), by means of an instrument (*ādhikaraṇikī*), produced by animosity (*prādveṣikī*), pain-causing (*pāritāpanikī*), and those that cause death to another being (*prāṇātipātikī*). See editor's note to the edition published in Beawar (Rajasthan), part 2, p. 574 (*Vyākhyāprajñapti Sūtra*, 4 vols., Jināgama Granthamālā, nos. 14, 18, 22, 25 [Beawar, Rajasthan: Sri Agam Prakashan Samiti, 1982–1986]).

35. See note 33 above.

36. *BhSū śataka* 1, *uddeśaka* 2, *sūtra* 7.2, p. 19 (39a), as translated by Lalwani, *Bhagavatī Sūtra*, 1:39–40.

37. Lalwani, *Bhagavatī Sūtra*, 1:239–40. See also Deleu, *Viyāhapannatti (Bhagavaī)*, 76.

38. *BhSū śataka* 19, *uddeśaka* 3, *sūtra* 33, p. 840 (766b), as translated by Deleu, *Viyāhapannatti (Bhagavaī)*, 250.

39. *Ācārāṅga Sūtra* 1.1.

40. Sylvan, "Is There a Need for a New, an Environmental, Ethic?" 20.

41. As defined in *Merriam-Webster's Collegiate Dictionary*, 10th ed., s.v.

42. As quoted by Goodpaster, "On Being Morally Considerable," 64.

43. *GJK* 134–39. Using the desire for food as an example, each is produced by external causes (e.g., the sight of food), which causes the premature realization (*udīraṇā*) of a specific *karma* (e.g., *asātā-vedanīya karma*). See also *Prajñāpanā Sūtra*, chap. 8.

44. Goodpaster, "On Being Morally Considerable," 65.

45. Callicott, "Introduction to Part One: Environmental Ethics," 11.

46. J. L. Jaini, trans., *Gommaṭasāra Jīvakāṇḍa*, 92: "Injury (*hiṃsā*) is defined as the deprivation through carelessness (*pramāda*) of any one or more of the vitalities (*prāṇas*) of a soul. Really speaking it is the vitalities to which injury is caused, nei-

ther the soul nor the matter which encases a mundane soul is susceptible to any injury whatsoever. They are both indestructible in their essence."

47. *Ācārāṅga Sūtra* 1.2.3.4, as translated by Jacobi, *Jaina Sūtras*, 19.

48. *Ācārāṅga Sūtra* 1.4.1.1, as translated by Jacobi, *Jaina Sūtras*, 36.

49. Paul W. Taylor, "The Ethics of Respect for Nature," in *Environmental Philosophy*, ed. Zimmerman et al.,79; originally published in *Environmental Ethics* 3, no. 3 (fall 1981): 197–218.

50. For a discussion of the appropriate conditions for beginning this process of spiritual awakening, see P. S. Jaini, *The Jaina Path of Purification*, 140–41. According to Jain tradition, in those locations of the universe that are subject to ascending and descending cycles of time (*utsarpiṇī* and *avasarpiṇī*), such as Bharata-kṣetra, where we are said to live, *mokṣa* is only possible during certain periods of time. The last person to attain *mokṣa* here in the current *avasarpiṇī* was Jambū, a disciple of the twenty-fourth Tīrthaṅkara, Mahāvīra. According to tradition, he attained *mokṣa* some sixty-four years after the death of Mahāvīra, whose death is often cited as 527 B.C.E. (see P. S. Jaini, *The Jaina Path of Purification*, 37 and 46). However, elsewhere in the universe (in the Videhas), the possibility of a human attaining *mokṣa* always exists because Tīrthaṅkaras are always preaching there. For a discussion of the Jain universe and the twenty-four Tīrthaṅkaras, or "Ford-builders," who are born at certain times in these cycles to lead others to *mokṣa*, see P. S. Jaini, *The Jaina Path of Purification*, 29–34.

51. See, for example, the story of the elephant Marubhūti (who was later born as the twenty-third Tīrthaṅkara, Pārśvanātha) in the *Triṣaṣṭiśalākāpuruṣacaritra* of Hemacandra, as translated by Helen Johnson (*The Lives of Sixty-three Illustrious Persons*, 6 vols. [Baroda: Oriental Institute, 1962], vol. 5, pp. 360 ff.). See also Christopher Key Chapple, *Nonviolence to Animals, Earth, and Self in Asian Traditions* (Albany: State University of New York Press, 1993); and Padmanabh S. Jaini, "Indian Perspectives on the Spirituality of Animals," in *Buddhist Philosophy and Culture: Essays in Honour of N. A. Jayawickrema*, ed. David J. Kalupahana and W. G. Weeraratne, 169–78 (Colombo: N. A. Jayawickrema Felicitation Volume Committee, 1987).

52. For a study of various types of animal homes in India, including *pinjrapoles*, see Deryck O. Lodrick, *Sacred Cows, Sacred Places: Origins and Survivals of Animal Homes in India* (Berkeley and Los Angeles: University of California Press, 1981).

53. See note 46 above.

54. P. S. Jaini, *The Jaina Path of Purification*, 242–43.

55. Ibid., 243.

56. The *Airyāpathikī Sūtra in Yoga Śāstra* 3.124, as translated by R. Williams, *Jaina Yoga: A Survey of Mediaeval Śrāvakācāras* (London: Oxford University Press. 1963), 203–4.

57. *Pratikramaṇa-sūtra* 49, as translated by Williams, *Jaina Yoga*, 207.

58. Harold Coward, "The Ecological Implications of Karma Theory," in *Purifying the Earthly Body of God: Religion and Ecology in Hindu India*, ed. Lance E. Nelson (Albany: State University of New York Press, 1998), 39–40.

59. Goodpaster, "On Being Morally Considerable," 59. As noted by Callicott ("Introduction to Part One: Environmental Ethics," 11), Goodpaster "expressly avoids the issue of how much weight we ought to give the interests of plants and other barely living beings." See also Goodpaster, "On Being Morally Considerable," 69.

60. Singer, "All Animals Are Equal," 31.

61. As translated by Lalwani, *Bhagavatī Sūtra*, 1:37.

62. J. L. Jaini, *Gommaṭasāra Jīvakāṇḍa*, 92.

63. *Environmental Ethics*, ed. Robert Elliot (Oxford: Oxford University Press, 1995), 15.

64. *GJK* 183 and commentary, as translated by J. L. Jaini, *Gommaṭasāra Jīvakāṇḍa*, 114–15.

65. *GJK* 184, as translated by J. L. Jaini, *Gommaṭasāra Jīvakāṇḍa*, 115.

66. See the chapter by Paul Dundas in this volume.

67. *Ādipurāṇa* 170, *pārva* 40.

68. For a description of the objects provided by the ten types of wish-fulfilling trees, see Williams, *Jaina Yoga*, 255.

69. For a study of these texts, see ibid.

70. Ibid., 121 quoting from *Sāgāradharmāmṛta*. According to P. S. Jaini (*The Jaina Path of Purification*, 80), this thirteenth-century work is the only such text known to have been written by a layperson. See also Williams, *Jaina Yoga*, 26–27.

71. Williams, *Jaina Yoga*, 122: *khaṇḍanī peṣaṇī cullī uda-kumbhaḥ pramārjanī pañca-sūnā gṛhasthasya tena mokṣaṃ na gacchati.*

72. Ibid., 100.

73. Coward, "The Ecological Implications of Karma Theory," 42.

74. Callicott (paraphrasing Goodpaster), "Introduction to Part One: Environmental Ethics," 11.

75. See Williams, *Jaina Yoga*, 122.

Challenges to the Possibility of a Jain Environmental Ethic

Green Jainism? Notes and Queries toward a Possible Jain Environmental Ethic

JOHN E. CORT

The environment is in crisis. This is not, however, a "natural" crisis. It has been caused by the actions of one species: humanity. The net result of centuries of human impact—and, in particular, of the past several centuries of rapid population and economic growth—has been a global ecosystem that is increasingly damaged and increasingly inhospitable to many forms of life. With few exceptions, humanity the world over has viewed itself as separate from and superior to nature. In one sense, therefore, the answer to this humanly caused environmental crisis is simple: we must establish (or reestablish) healthy connections between humanity and nature and recognize that our very survival as a species depends, in the words of the American farmer, environmentalist, and author Wendell Berry, on "getting along with nature."[1]

How are we to establish and reestablish these connections? How are we to unlearn patterns established by the global industrial economy and learn (and relearn) patterns that support an ethic conducive to the flourishing of both humanity and nature? If a defining characteristic of humanity as a species is culture, then both the causes of and answers to our destructive behavior will be found in culture. As Berry observes, "we have only two sources of instruction: nature herself and our cultural tradition."[2] Paying close attention to nature will teach us much that is essential. This knowledge must be complemented by a (re)discovery of cultural traditions that allow us as humans to live in continuity with rather than in opposition to nature. Since religion is one of the core constituents of culture, the answers will involve a (re)discovery of our religious traditions for values and practices that support an ecologically positive way of living and an interrogation of our religious traditions to see where they promote values and practices that are harmful to the environment. For Berry, as an American of European descent, these cultural and religious traditions are what he terms "the Greek and Biblical lineages of our culture."[3] These two lineages are not, however, the only sources of vital cultural knowledge, and they may well be inappropriate for most people. For Jains, whether living in India or among the communities that have recently migrated to all parts of the globe, learning from nature how to live in continuity with culture will be complemented by learning from their own cultural traditions, more broadly South Asian and more specifically Jain. These traditions are no longer alone in shaping Jain life; but, for many Jains, they are crucial in the task of developing responses to the environmental crisis.

A goal of the conference from which this volume derives, as expressed in the invitation letter, was to "explore how traditional Jaina ethics, cosmology, and metaphysics might contribute to the emerging field of environmental philosophy."[4] In other words, the conference organizers located themselves within the field of environmental philosophy and were interested to hear what traditional Jain ideas, beliefs, and practices could contribute to their field. My approach in this essay—and it is precisely that, an *essay*, i.e., more speculative and suggestive than analytic or descriptive—is to reverse the direction of the information flow. While this essay will bring information from the Jain religious worldview and Jain lived experience to the scholarly world of environmental philosophy, this is an incidental rather than primary goal. Instead, I am addressing this to my many Jain friends and colleagues in India, Great Britain, and North America, who are concerned to respond to the environmental crisis and who over the years have asked me as a scholar of Jainism what I think might be distinctive about a Jain response to environmental issues. Accordingly, I have two intended goals. The first is to indicate to these engaged Jain[5] friends some of the questions, assumptions, and practices that they might want to think about and act upon as they develop a Jain environmental ethic. The second is to point tentatively to some ways in which Jains might begin to respond to these questions, by pointing to Jain practices and values that might well underlie a Jain environmental ethic or be adapted for such an ethic. There is much here that will not be new to colleagues in environmental studies but will, I assume, be useful to Jains who read this volume.

In January 2000 I was invited to participate in a conference in Jaipur to celebrate the one hundredth anniversary of the Digambara scholar and social reformer Paṇḍit Cainsukhdās Nyāytīrth. The day I attended was devoted to the conception of the environment in Sanskrit and Prakrit literature (Saṃskṛt evaṃ Prākṛt Sāhitya mẽ Paryāvaraṇ kī Avadhārṇā). More than a dozen papers were delivered in Hindi by Digambara scholars from all over North India.[6] Some of the papers fit into the field of environmental history, as they explored how nature and the environment have been portrayed in classical Jain literature. The majority of the papers were more apologetic in nature, as they surveyed various aspects of Jain doctrine to advance the position that Jainism is an inherently environmental religious tradition. What was striking to me was the disjunction between these papers, and the larger vernacular discourse on Jainism and the environment of which they are part,[7] and the extensive English-language scholarship on environmentalism published in India. One does not find any references to the vernacular and English scholarships in each other, and the two appear to exist in parallel, unrelated intellectual universes.[8] This essay intends to indicate some of the issues that need to be considered to bring these two discourses into dialogue with each other.

Ecology: A New Global Episteme

Let me start by stating an assumption that underlies this essay. To put it boldly, as of the early 2000s there is no Jain environmental ethic per se. Statements that Jainism is an inherently environmental religious tradition or that Jainism has always "enthroned the philosophy of ecological harmony" are largely untrue as statements about history, and I would argue that such mis-statements will hinder more than help in the development of a Jain environmental ethic.[9]

The reason I say that there is no Jain environmental ethic is that environmentalism is a relatively new episteme worldwide. It has arisen out of a set of physical, technological, and increasingly moral and intellectual challenges of the past several centuries, but has attained its position as a distinct field of inquiry—an episteme—only within the past several decades. In the words of Harold Coward, "It is only recently that the various religions have had to question their sources with regard to the interaction of humans with the environment—in response to the explosion in numbers of people and their consumption of the earth's resources at a rate that threatens to exhaust its life sustaining capacity."[10] Environmentalism raises a new set of questions and issues, hitherto not addressed explicitly by Jains either in practice or in thought. This is not to say that Jains have not thought about and acted toward the environment. But the concept of "environment/ ecology/ nature" is a new episteme (the very fluidity of terms is indicative of its newness), raising questions and issues that Jains have not addressed in this particular formulation. To put it simply, without the episteme of environmentalism, it is not possible to develop a conscious response in thought, speech, and deed to the questions raised by that episteme. Hence, to speak of "Jain environmentalism" before the recent past is meaningless. This is only one of several new epistemes to which the world's religious traditions have had to respond in recent centuries; others include the scientific method, Copernican astronomy, nationalism, industrial capitalism, globalization, feminism, social justice, human rights, nuclear weapons, and cultural and religious pluralism, to name a few.[11]

Any Jain environmental ethic at present is at best nascent and largely unconscious and implicit. Engaged Jains are just beginning the task of articulating such an ethic. To a significant extent these articulations are to be found in practices and habits rather than in systematized statements, for this is the way any lived ethic develops. But the external pressures of the environmental crisis, as well as the efforts of engaged intellectuals and theologians in other religious communities to develop their own explicit environmental ethics, have resulted in the beginning of efforts by Jains to explore their own tradition within the new epistemological framework of environmentalism.

Steps toward Developing a Jain Environmental Ethic

There are three major aspects of this process of developing a Jain environmental ethic. These are not sequential steps but, rather, three different kinds of exploration. While of necessity they will occur simultaneously, I discuss them separately for the sake of clarity.

The first task is historical, for any ethic is based in significant part on the historical particularities of a community. This is a largely descriptive process of the investigation and documentation of Jain understandings and practices that indicate how Jains have understood nature and the place of humanity within nature. This investigation will involve seeing how Jains have both explicitly and implicitly defined "nature," for the definitions of this term are themselves culturally located. In other words, "nature" itself is a cultural category, and so is defined differently in different cultural worldviews. Part of this process, obviously, is the philological one of exploring the meanings and contexts of the variety of Indic words used to refer to the range of referents of the English word "nature." In English, "nature" has two partially contrasting meanings—as the totality of the physical universe, and as that which is apart from humanity, civilization, and culture—which indicate how humanity is at once a part of and apart from nature.[12] The study of Jain attitudes toward the environment will involve a careful study of the semantic fields within Jain thought and practice of a number of overlapping terms, such as *prākṛti*, *lokākāśa*, *dravya*, *ajīva*, *svabhāva*, and *paryāvaraṇa*.

Several of the papers in this volume are engaged in this process of historical investigation. In part, this will involve an investigation of Jain ontology and metaphysics, looking at Jain conceptions of matter, soul, and the like. But I would argue that this alone would be inadequate. To understand how Jains have understood nature, and how those understandings have shaped Jain lived experience, there are better sources than philosophical doctrines. In particular, it is here that the skills of cultural interpretation and literary and artistic analysis can be brought to bear to investigate what Jain narratives, myths, histories, paintings, sculptures, and practices can tell us about Jain assumptions about the nature of nature.[13]

This investigation into an environmental history of Jainism needs to be grounded in the specifics of Jain cultures and societies. Rather than look at material only at the abstract and reified level of "Jains" and "Jainism," it should also recognize the developments in recent Jain studies that have focused on the specific expressions of the four sectarian traditions, different regional traditions, and the ways these have changed over time.[14] Understandings of and reactions to the environment are always *local*. Scholarship on the environmental history of Jainism therefore should focus on cultures and practices as localized, and localized Jainism is always sectarian.[15]

For those who are concerned with the environment, this historical exploration is often an unhappy process, as the historical record of the Jains is, on the whole, not a positive one. The Jains are not alone in this, as I doubt that any religious tradition has a salubrious record on matters of environmental and interspecies justice. But it is an important task, for without understanding the practices and attitudes that have shaped past actions, it is much more difficult to develop creative responses in the present.[16] The importance of recognizing and rectifying past wrongs is at the very heart of Jain praxis, for the speaking of the truth (*satya*) has a prominent place among the vows (*mahāvrata*s) taken by Jain mendicants.[17] Among the amplifications of this vow is that it applies equally to the past, present, and future. I interpret this as a call for Jains to be forthrightly honest about blemishes in their own history. This is also recognized in the ritual of *pratikramaṇa* as practiced by both mendicants and laity, in which the individual acknowledges past transgressions and seeks to rectify the negative karmic results of those transgressions; speaking falsely (*asatya*) ranks right after harm (*hiṃsā*) in the standard list of the eighteen most serious transgressions.[18]

The second task is also to a significant extent historical. This is the programmatic recovery of those narratives and practices that can serve as bases for a Jain environmental ethic. Some of these will be narratives and practices that come from mainstream traditions, known and practiced by many Jains, but now understood within a new framework. Others will be alternative lineages and voices, people and events from Jain history that have not necessarily been part of the mainstream traditions, but which can provide a historical grounding for a Jain environmentalism.[19] This recovery scholarship, however, needs to remain intellectually honest. It must always remember that the goal is to uncover what Jains *can do*, *might do*, or *should do* in the future, based on these historical models; this is quite different from the language of many contemporary Jain enthusiasts who mistakenly convert these models into assertions that they represent what Jains *have done*. Recovery scholarship will play an important role in developing Jain environmental practices, for narratives are more likely to inspire people to action than are abstract philosophical principles. Myths and other narratives provide us with stories that both allow us to make sense of our world and to see how we can change our world. Gary Nabhan has perceptively discussed the transformative power of narratives:

> To restore any place, we must also begin to re-story it, to make it a lesson of our legends, festivals, and seasonal rites. Story is the way we encode deep-seated values within our culture. Ritual is the way we enact them. . . . By replenishing the land with our stories, we let the wild voices around us guide the restoration work we do. The stories will outlast us.[20]

The third task is one of action and reflection in the present. This is the work of conversation and interaction between engaged Jains and non-Jain environmental activists and theorists, in which the principles, practices, and worldview assumptions of Jainism are placed in a condition of dynamic and potentially fruitful interchange with the principles, practices, and worldview assumptions of environmentalists. In some ways this is precisely the goal behind the series of conferences out of which this volume emerges. Perhaps more importantly, it is also what happens when Jains are actively involved in environmental issues and campaigns in their neighborhoods, workplaces, schools, and broader communities. This task involves a creative interaction between Jain ways of being and currents of contemporary environmental thought. This should be a mutually cross-fertilizing process. It is not a matter of mapping environmental principles onto Jainism, nor one of mapping Jain principles onto environmental practice, but of exploring aspects of Jain thought and practice that will form the bases for a distinctively Jain environmental ethic.

In the rest of this essay, I indicate some of the issues and practices that may well come into focus from Jain interaction with environmentalism. To frame this discussion I introduce some of the major issues that have emerged in environmental thought and practice, both North and South, in recent decades. I express these issues in the form of questions that the new episteme of environmentalism addresses to Jains. There are no obvious answers to these questions, but I think they merit careful consideration by Jains, for it is largely in their answers, as expressed in thought, speech, and action, that Jains will develop their own distinctive environmental ethic.

Liberation from the World, Wellbeing in the World

Before raising these queries, I should mention one other key assumption that informs my discussion, an assumption of where I think that Jains can fruitfully look within their own traditions for responses to the episteme of environmentalism. Jainism is frequently presented both by Jain intellectuals and scholars of Jainism as a distinctive path to liberation (*mokṣa-mārga*), an intertwined set of doctrines, practices, and worldview assumptions focusing on the ultimate liberation of the human soul from bondage.[21] The *mokṣa-mārga* ideology is not very conducive to the development of an environmental ethic. At its heart is the goal of permanent separation of the soul from all matter. In such a dualist ideology positive environmental results are largely incidental.[22] This body-spirit dualism is found embedded in a wide variety of doctrines, beliefs, and practices within the Jain tradition, from the many ways in which Jains "filter" the material world in pursuit of a purer spirituality, to ascetic dietary practices that in their extreme cases can be interpreted as expressing a fear of the biological world, to the Digambara mystical emphasis on the need for an existential realization of the ultimate difference (*bheda-jñāna*) between soul and matter.[23]

But the *mokṣa-mārga* ideology is not the whole of Jainism, for Jainism also is a religious culture that provides people with a definition of a good life in this lifetime, what I have elsewhere termed the value of "wellbeing." The realm of wellbeing involves a much less negative (albeit still not unreservedly positive) attitude toward the nonhuman world, toward the physical world, and toward our own physical embodiedness.[24]

Looking at the actions and beliefs expressive of wellbeing will be a much more fruitful avenue for developing a Jain environmental ethic than looking at the actions, beliefs, and doctrines expressive of the *mokṣa-mārga*.[25] It is important to remember that the *mokṣa-mārga* ideology itself presents Jainism as a graduated path. Looking only to the *mokṣa-mārga* ideology results in focusing too exclusively on the higher, more rarified rungs of the path. These rungs are almost impossible for a human to attain, and so establish a set of unrealistic goals for environmentally concerned Jains. This approach also downplays the sociobiological contexts in which Jains live and in which any Jain environmental praxis will be located.

One other aspect of the *mokṣa-mārga* ideology itself leads to the conclusion that we must not privilege it as the totality of Jainism when searching for the bases of an environmental ethic. The Jain community as created by the Tīrthaṅkaras consists of four *tīrtha*s (social subdivisions), not just two. There are the *tīrtha*s of the *sādhu* and the *sādhvī*, the male and female mendicants whose practice is based on world renunciation and the focused pursuit of *mokṣa*. But, equally important within the Jain community are the *tīrtha*s of the *śrāvaka* and *śrāvikā*, the laymen and laywomen, whose practice is *not* based solely on renunciation. Jainism cannot be reduced to just world renunciation in all its many forms. Jainism also involves responsible, moral action, including action concerning and within the environment.

European-American ethical systems tend to be phrased in terms of universals and so are expected to be followed equally by all people at all times. Jain ethics (and South Asian ethical systems more generally), on the other hand, are highly context-sensitive.[26] Full-fledged mendicants, who have taken the great vows, are expected to observe Jain ethical principles universally. In the technical language of Jain praxis, the vows are *sarvavirati*, or universally applicable. But, for the laity, who make up the vast majority of Jains, such strenuous observance is considered impractical if not impossible, and so lay observance is *deśavirati*, or in accordance with one's socio-moral location. Any deed, thought, or word is judged according to three factors: the location (*deśa*), the time (*kāla*), and the actors (*pātra*). This provides an essential flexibility to lay ethics and calls for the creative response to ethical dilemmas rather than an unthinking application of ethical a priori. Such an approach, in my opinion, is essential for the solution of environmental problems, which always involves the balancing of the multiple, conflicting needs and aspirations of many beings, both human and nonhuman.

Queries for Environmentally Engaged Jains

Let me now turn to what I perceive to be some of the important questions that the new episteme of environmentalism poses for Jains. As I said above, neither the questions themselves nor the answers that have developed within Western religious and secular traditions will be new to most of my colleagues within environmental studies. But I suspect that they have only been indirectly apprehended by most of the Jains engaged in developing an environmental Jainism. Raising these questions to a level of conscious awareness may result in these engaged Jains being more self-reflexive as they create a Jain environmental ethic and may also allow them to avoid the danger of coming to answers and conclusions at a premature stage of understanding.[27] Some of these are questions that are foundational to all environmental thought and practice. Others are questions raised by specific movements and fields within environmental thought and practice, such as deep ecology, environmental justice, and ecofeminism. These different sub-epistemes within environmental thought (and there are, of course, more than just those I will address here) each address a different set of queries to Jainism, for each starts from its own set of presuppositions, questions, and concerns.[28]

Bodies: Human, Nonhuman, Divine

The most basic question that arises from the interaction between environmental thought and any religious tradition concerns that tradition's definitions and presumptions in response to an interlocking set of cosmological and ontological questions. How does the tradition define a human being? How is a human defined, both as an individual and as a member of larger human collectivities? In what ways is a human the same as and different from all that is characterized as nonhuman? How does it understand the human body? How does it understand and define the nonhuman?

Such definitions frequently revolve around distinctions between human and nonhuman or between culture and nature. But the very phrase "human nature" indicates that such distinctions are rarely, if ever, watertight. Rather, when we look at the evidence of both systematic reflection and lived experience, we usually find an interrelated and sometimes mutually inconsistent net of understandings and pre-understandings, expressed in thought, word, and action. Further, we find that in some instances the key distinctions are between human and nonhuman domains, while in other instances they are between body and soul (or mind) or between matter and spirit.

Jain responses to these questions are found most clearly in discussions of the fundamental ontological categories (*dravya* or *astikāya*), variously counted as five (soul, motion, rest, atoms, and space) or six (adding time).[29] These can be reduced to a fundamental dualism between sentient soul (*jīva*) and all that is insentient, including matter (*ajīva*).

This dualism, which is rooted in the *mokṣa-mārga* ideology, is most unhelpful for developing a Jain environmental ethic. The soteriological goal according to this ideology is for each soul to achieve liberation from all that is not soul, thus establishing a clear devaluation of nature. But, as I indicated above, the upper reaches of the *mokṣa-mārga* ideology are not where Jains should look to develop an environmental ethic. A more useful approach here is to look at Jain biology, which forms the basis of much Jain practice.

The Jain worldview posits the near ubiquity of souls in the universe. Each of these souls in its ideal form is identical in its qualities of bliss, energy, and omniscience, but due to each soul's unique karmic history, these souls are embodied in various forms. Jain biology distinguishes these forms in terms of different kinds of body (*kāya*), ranging from those with five senses through those with just the single sense of touch.[30] Here, I think it is significant that Jain biology distinguishes these embodied states in terms of *kāya*, or body: the concern is not just for the soul that is embodied, but also for the body itself. This focus on the many possible embodied states of the soul can help mitigate the dualism entailed in the soul–nonsoul distinction. Further, the Jain understanding of the possible range of bodies also extends beyond that found in some other worldviews, for bodies include not only animals and plants, but also such forms as air, water, and earth.

The strand of environmental thought known as deep ecology[31] here asks about Jainism's ethical valuation of this diversity of life-forms. What is the operative context that frames a Jain environmental ethic? Is it framed primarily around human concerns, or does it give some value, or even equal value, to the nonhuman? Is diversity valued because it provides additional resources for human existence and development, or is diversity understood to have intrinsic value?[32] These questions are amplified by that branch of environmental ethics which has attempted to develop an ecocentric, nonanthropocentric, environmental ethic.[33] Ecocentric ethical thought prompts us to ask of the Jains, what are the rights of the various nonhuman bodies? Such a question, however, is embedded in a number of religio-cultural presuppositions, for the very language of rights is based upon conceptions of justice with their triune roots in Abrahamic covenantal theology, Greco-Roman legal ethics, and the Enlightenment understanding of the person as an autonomous and rational individual. Certainly, the question of the rights (if any) of nonhuman bodies is one that Jain environmentalists will want to investigate. Jains do not usually frame the matter in terms of the rights of the nonhuman. Instead, they speak of the moral responsibility of humans toward all bodies, a responsibility most clearly enunciated in the cardinal Jain moral principle of *ahiṃsā*, or nonharm.

Several authors in this volume discuss the role of *ahiṃsā* in Jain environmental ethics, so I will restrict myself to just a few comments here. In the understanding of *ahiṃsā* found in the vows (*mahāvratas*) of the mendicants, its observance is said to be thrice threefold. The practice of *ahiṃsā* involves mind, body, and speech, and so is a matter of intention as much as action. Second, it involves what we ourselves think, do, and say, but also what we have others think, do, and say, and the public approval (*anumodana*) or censure (*nindā*) of the thoughts, actions, and words of others. Introducing *anumodana* and *nindā* into the understanding of *ahiṃsā* means that Jains are expected to be interventionist in their ethics. To stand by idly while someone else acts in a way that is harmful to the environment involves the bystander as much as the actor in the moral harm of the deed. Third, a Jain is expected to observe *ahiṃsā* in the past, present, and future. This means that one has a moral responsibility for prior harmful actions, both one's own and those of others. It also means that one has a moral responsibility to the future. Allowing environmental degradation today is not only violence in the present, it is also violence in the future. While *ahiṃsā* has not traditionally been expressed in terms of rights, this understanding of one's moral responsibility for *ahiṃsā* in the future might bear fruitful comparison with discussions by ecocentric ethicists on the rights of future beings to be born into a healthy environment.

When one juxtaposes *ahiṃsā* with the Jain understanding of the *kāyas*, one sees that *ahiṃsā* is not merely a matter of not harming one's fellow human beings. Jains have almost universally understood *ahiṃsā* to entail being vegetarian, and the unique ways in which *ahiṃsā* has informed Jain diet are well known.[34] The full application of *ahiṃsā*, however, involves applying it not just to the gross, obvious forms of life, such as humans and five-sensed animals, but also to the very essential prerequisites of life—to the air, water, earth, and plants—

for, according to Jain biology, all of these serve as the abodes of countless souls.

This understanding of *ahiṃsā* as applying to the fullest range of bodies is found in the rite of *pratikramaṇa*, performed twice daily by Śvetāmbara mendicants and, ideally, at least once a year by laity. The rite begins with the individual recognizing and seeking to absolve himself or herself from the karmic consequences of any form of harm caused to a wide array of life-forms, in bodies with from one to five senses, including seeds, plants, dew, insects, mold, and spiders. As a further part of this rite, the individual recites the following Prakritized vernacular liturgy (here given in its Tapā Gaccha form), in which he or she enumerates all the possible bodily forms, with the assumption that one has wittingly or unwittingly harmed them:

> 700,000 earth bodies,
> 700,000 water bodies,
> 700,000 fire bodies,
> 700,000 air bodies,
> 1,000,000 separate plant bodies,
> 1,400,000 aggregated plant bodies,[35]
> 200,000 two-sensed beings,
> 200,000 three-sensed beings,
> 200,000 four-sensed beings,
> 400,000 divine five-sensed beings,
> 400,000 infernal five-sensed beings,
> 400,000 plant-and-animal five-sensed beings,[36]
> 1,400,000 humans:
> in this way there are 8,400,000 forms of existence.
> Whatever harm I have done,
> caused to be done,
> or approved of,
> by mind,
> speech,
> or body,
> against all of them:
> may that harm be without consequence.[37]

Most of the *pratikramaṇa* rite is recited in Prakrit or Sanskrit. But this particular part of the liturgy is recited in the vernacular, indicating it is intended to be clearly understood by the practitioner. I would suggest that this rite of confession could be creatively adopted by Jains as an environmental ritual.[38]

To complicate further the question of how a religious tradition understands the relationships between human and nonhuman, between soul and body, we must add a third factor, that of theology: how are both the human and the nonhuman, the soul and the body, understood in relation to the superhuman or divine? This is another issue that warrants much greater attention than I will give it here. But let me again make a few passing remarks. Jainism is distinctive (although by no means unique) among the world's religious traditions in its vigorous denial of any definition of God that posits God as creator of the material and/or spiritual universe. Both soul and matter, according to the Jain worldview, have existed from beginningless time and will exist into the endless future. Jains worship as God all souls that have liberated themselves from karmic bondage.[39] While there is a crucial difference between the liberated and unliberated souls, in that the former have transcended the realm of the *kāyas*, there is no difference in terms of the ultimate ontology of liberated and unliberated souls. I am simplifying a very complex issue here. But this is a matter which bears further reflection, for central to any religious tradition's environmental ethic is its understanding of God (or other sacred ultimate) and of the relationships among God, humanity, and nonhuman nature.[40] The Jain understanding of the ubiquity and uniformity of soul among all three estates (human, natural, divine) will result in a different environmental ethic from traditions such as the three Abrahamic ones, which are based on the crucial distinction between God and humanity.[41]

Jain Women and Ecology

The branch of environmentalism known as ecofeminism further expands upon the scope of questions concerning the understandings of the relationships and interactions among the divine, human, and nonhuman natural realms.[42] Ecofeminist thought and practice starts from the recognition that the lived experiences of men and women are quite different. Further, ecofeminism starts from the recognition that gender differences themselves are as culturally constructed as the differences between nature and culture, and so calls our attention to the powerful role of such constructs as "Mother Nature" and "Earth Goddess" (Bhūdevī) and the distinction between "nature" (*prākṛti*) as feminine and "intellect" (*puruṣa*) as masculine in shaping our experience of reality.[43] Third, ecofeminism asks us to consider the connections between the human oppression of nature and other forms of oppression, in particular the patriarchal oppression of women. Are there connections between "nature hating" and "woman hating"[44] in Jain doctrine and practice? Since ecofeminism is as much concerned with action in response to perceived injustices as it is with analysis of the causes of those injustices, it also asks us to consider if the response to one form of oppression might not be related to responses to other forms of oppression. If, as ecofeminism posits, the androcentric oppression of women by men and the anthropocentric oppression of nature by humans are expressions of the same (or related) hierarchical expressions of oppressive power, then might the redress of one form of oppression be linked to the redress of the other?

Scholarship has only recently begun to explore the ways in which the Jain tradition has both shaped gendered experience and been shaped by the different gendered experiences of men and women.[45] In many respects, Jain values are no different here from those of the broader values encompassing South Asian cultures. While the exact soteriological abilities of women have been debated for many centuries,[46] many Jain texts of all sectarian traditions are highly gynophobic. Jain monastic ideology has almost always ranked all monks above all nuns, and gender hierarchy has been pervasive throughout Jain society. At the same time, Jain monastic traditions have been notable in South Asia for the large number of nuns, who in many times and places have outnumbered the monks severalfold. Women have played central roles in the preservation and reproduction of Jain religious culture, and they have had their own religious spaces within the tradition.[47] How these and other factors interact with Jain attitudes toward the environment is an open and investigable question; both scholarly research and engaged Jain reflection may well reveal distinctively Jain expressions at the intersections of ecology and gender and environmentalism and feminism.

Jains and Environmental Justice

The questions posed of any Jain environmental ethic by ecofeminism bring the issue of power squarely into the picture. Similar questions are also raised by two different but related streams of environmentalism, known generally as environmental justice and social ecology (based in the industrialized North) and Southern environmentalism (based in the developing South).[48] In this field ecological advocates stress that any strong environmental ethic must be built equally on diversity, sustainability, and equity.[49] In developing what he terms an "environmentalism of the poor," the Indian environmentalist Ramachandra Guha emphasizes that in many parts of the world, especially in the South but also in large pockets of the North, "for the sections of society most critically affected by environmental degradation—poor and landless peasants, women, and tribals—it is a question of sheer survival, not enhancing the quality of life. . . . [A]s a consequence, the environmental solutions they articulate strongly involve questions of equity as well as economic and political redistribution."[50]

Environmental justice therefore posits that issues of environmental and interspecies justice cannot be separated from issues of social and intraspecies justice. Just as nonhuman species have rights to live, so do all humans have the right to a safe, secure, and sustainable livelihood. Environmental justice recognizes that there is a material basis to the environmental crisis, that it is in part a matter of control over and distribution of resources, as well as control over the negative environmental effects of industrial production. As an example, Robert D. Bullard has shown how toxic waste dumps in the United States are disproportionately located among communities of racial and ethnic minorities, as well as others who are economically and politically dispossessed.[51] Part of the answer to the environmental crisis, therefore, according to environmental justice, lies in addressing issues of inequity and injustice in the control and distribution of resources.[52]

Environmental justice therefore asks of the Jain tradition, what is the place of social equity within human interaction? Environmental justice activists and thinkers have argued that actions to protect the environment at the expense of people who depend upon that environment for their livelihood lead to social injustice and, in the long run, undermine those actions.

Here, the example of the attempts to reforest the sacred Mūrtipūjaka Jain mountain of Śatruñjaya are illustrative. Medieval accounts describe it as forested, but the contemporary experience is of a denuded mountain. In recent years there has been an effort to reforest the mountain, aided by significant contributions from overseas Jains.

Anyone who has had the experience of walking up both a forested and a deforested mountain would think that this effort is unquestionably laudable.

But, what is the cause of the mountain's deforestation? It is not the result of any natural catastrophe or climactic change. Rather, the trees and bushes have gradually disappeared due to local lower-caste herders grazing their cattle and goats. In response to the pressures of this pastoral economy, the governmental authorities and the upper-caste Jains engaged in the reforestation project have erected fences of thorn bushes and adopted other methods to deprive the herders of their grazing lands, rights which have existed for centuries. Here we see an effort to improve the environment, but at the expense of those who are poorest and most dependent upon the resources, and who have not been involved in either the decision-making or implementation processes. The laudable ecological effort to reforest the mountain has dispossessed the poorer herders of an important means of livelihood.[53]

Environmental justice thus insists that the rights of nature must be balanced by the rights of humans, especially those who are the poorest. It raises the question of human consumption patterns and resource sustainability. While overpopulation itself is an environmental problem in many parts of the world, the unequal consumption of resources is a greater problem. In this the Jains find themselves in the position of being for the most part residents of the South, but successfully aspiring to the consumption patterns of the elites in the North. Can Jains balance their social location, as members of the Southern elite, many of whom are also striving to enter the upper echelons of the North, with the environmental need to redress patterns of consumption? Can Jains find ways to make their involvement with the world not one in which they monopolize resources for themselves and externalize the social and environmental costs onto the poor, both Jain and non-Jain? Can they find ways to strive to mobilize Jain values in support of a more just distribution of resources?

Without addressing issues of economic justice, there can be no lasting and significant contribution to the environment's well-being.[54] During the twentieth century most Jains have wholeheartedly embraced the values of global industrial and postindustrial capitalism, and have thereby contributed significantly to environmental and social degradation. At the same time, one does find among both Indian and diaspora Jains alternative voices calling for attention to the needs of the poor and dispossessed. Can Jain activists amplify these latter voices and mobilize Jains on behalf of economic and environmental justice?

In this context environmental justice also asks, in what ways does the Jain tradition provide a response to the issue of overconsumption? On the one hand, anyone observing the wealth of the Jain community would assume that there is not a ready answer. But here I think that the Jain tradition does have values deeply embedded in both practice and thought that can be brought to the foreground. On many occasions when I have addressed groups of young Jains in North America, they themselves have commented on the paradox that the Jains on the whole are an aggressively accumulating community, while the value of nonpossession (*aparigraha*) is at the center of the stated ideals of both mendicant and lay life. Clearly, this is a paradox worthy of deeper consideration by Jain environmentalists.

Looking at some of the ideals expressed in the lay vows indicates ways in which the Jain tradition understands the problems of overconsumption. The vow of *asteya* is usually understood to mean, simply, not stealing, but in textual discussions it is more broadly understood to entail not taking anything that has not been freely given, whether by a person or by another living creature. This could easily be read to mean that many of the ways in which one accumulates resources within the industrial capitalist system, whether from nature or from other humans, are morally problematic. Similarly, two of the vows recommended for laity are also amenable to an environmental reading.[55] The vow of *bhogopabhoga-parimāṇa*, or enjoining the consumption of a number of forbidden items, is generally applied only to diet but could easily be applied to overconsumption more generally. The vow of *anarthadaṇḍa*, or enjoining a number of harmful occupations and activities, is also usually understood to refer to a narrow range of occupations that clearly violate *ahiṃsā*. But this also could easily be extended to involve reflection on the environmental and justice consequences of one's occupation and consumer patterns. Such reflection is already beginning among younger diaspora Jains. In May 1997 Young Jains, a British organization, held a conference entitled "Jainism in Business and Professional Life" in Watford, England. Atul K. Shah, one of the organizers of the conference, wrote:

> Our happiness is directly connected to the happiness of others. If we are rich and others are poor, our happiness will only be temporary.

> Sooner or later, the poor will knock at our door, and some may even knock it down. Just because they live far away in slums and we cannot see them does not mean that we will be able to ignore them for long. Jainism argues that we can never be rich when others are poor. True richness comes from sharing and giving, from raising ourselves and others at the same time. . . . How can we call ourselves successful, when the society to which we belong kills and plunders so indiscriminately? How can we call ourselves Jains when the firms we invest in are greedy, violent and destructive?[56]

Finally, some proponents of environmental justice have argued that the problem of environmental abuse, or human violence toward the nonhuman, cannot be separated from the problem of war and from all other forms of human violence against other humans. Again, this is an issue that would take us far afield, and so I will mention in passing that just as Jain attitudes and practices toward the environment are changing in the context of the new episteme of environmentalism, so Jain attitudes and practices of *ahiṃsā*, which have traditionally been expressed largely in terms of diet, are also facing queries from other traditions of nonviolence, such as those of religious and secular pacifism and nonviolent action, that root nonviolent action in understandings of social justice. Jains have rarely understood *ahiṃsā* as involving them in participation in movements to minimize violence in society at large or in efforts to resist war and militarization, but environmentalists ask if such a narrow compass for nonviolence should not be expanded.[57]

Ecology as Local and Regional

This leads me to a final question related to those raised concerning the connections between environmental and social justice. From among the many issues raised by the polycentric field known as deep ecology, let me turn to those raised by bioregionalism.[58] Bioregionalism asks what the proper relation of a human population is to its bioregion. Emerging out of this question, it further asks whether or not centralized social structures are inevitably oppressive of both humans and the environment.

Bioregionalism advocates decentralized, self-determined modes of social organization and culture that are predicated upon biological integrities as measured by what are termed bioregions. A bioregion is determined by three factors: 1) the biotic shift, or change in plant and animal species, is less than 15 to 25 percent; 2) it lies within an integral watershed and other landforms; and 3) it exhibits clear cultural continuities. Within any bioregion, there may be any of four different inhabitory zones, each of which entails distinctive modes of livelihood and interaction with the environment: cities, suburbs, rural areas, and wilderness. Bioregionalism, as with environmental justice, focuses on human consumption patterns; but, whereas environmental justice advocates are divided in their attitudes toward the possibility of unending economic growth, bioregionalists tend to assume its impossibility and so the need to develop sustainable, stable economic systems not predicated on growth. Bioregionalism is perhaps best exemplified by the oft-reprinted questionnaire entitled "Where You At?" which emphasizes basic cultural and environmental knowledge of one's bioregion.[59]

In many ways bioregionalism emerges from a distinctly North American (and Australian) cultural context of a highly urbanized society coupled with a large expanse of underpopulated and depopulated land. Thus, some of the questions raised by bioregionalism are inappropriate for people living in India. But others of the questions are still pertinent to India, and I would argue that many of them are especially pertinent for a community such as the Jains that has also been highly urbanized for many centuries.

Bioregionalism asks of the Jains, to what extent are environmental problems caused or exacerbated by Jain understandings of and attitudes toward their bioregions? Have Jains contributed to healthy environmental development within the cities and towns where they tend to live? The history of the Jain communities of northern and western India is a history of frequent migrations. Has this pattern contributed to a lack of environmental sensitivity to where they live? What about the environmental consequences of the semi-peripatetic patterns both enjoined upon Jain mendicants and often times necessary for Jain traders, or what about the North American patterns of residential mobility being adopted by many immigrant Jains?

Here, the requirement that all mendicants cease their peregrinations for the four months of the rainy season could be reinterpreted as a call for them to develop greater connections with specific bioregions. Similarly, two of the twelve vows recommended for laity—

dig and *deśāvakāśika*—involve the individual vowing not to go beyond a certain geographical limit, and so they could be reinterpreted in a bioregional light.

Some of the Jain mindfulness techniques could be interpreted as calling for greater attention to bioregion. *Sāmāyika*, the principle Jain technique of meditation, is understood in the case of mendicants to involve perpetual awareness of all of one's actions, both voluntary and involuntary, especially with an eye toward reducing occasions of causing harm to other bodies. It is also practiced by many laity for short periods on a regular basis. Clearly, this could be developed into a form of environmental mindfulness.

In a similar fashion, the five *samiti*s, or rules of conduct, that amplify the mendicant great vows[60] could be fruitfully applied to environmental awareness. Care in walking (*īryā-samiti*) could call for one to pay attention to the environmental consequences of all one's modes of transport. Care in accepting things (*eṣaṇā-samiti*) could be expanded to entail considering the environmental history of objects that come into one's life, from modes of extraction and production to modes of transportation, marketing, and selling. Care in picking up and putting down things (*ādāna-nikṣepaṇa-samiti*) clearly calls on one to pay attention to one's surroundings. Finally, care in the performance of excretory functions (*utsarga-samiti*) calls on one to investigate what happens both to waste items that one disposes of personally and waste that is a by-product of extractive and manufacturing processes.

Concluding Observations

In this essay I have attempted to indicate what I perceive to be some of the questions that the new and developing episteme of environmental thought and practice poses for Jains. I have also indicated some of the ways in which Jains might creatively investigate and reinterpret their own traditional modes of thought, speech, and action as they strive to develop distinctively and authentically Jain responses to the global environmental crisis.

The Jain emphasis on adherence to the truth in past, present, and future will mean that Jains need to look unblinkingly at the many ways they have wittingly and unwittingly contributed to environmental degradation. Such an adherence to truth can also be a most powerful tool for social change; this can be seen in both the Gandhian emphasis on *satyāgraha*, or nonviolent action in pursuit of the truth, and the Quaker assertion that struggling for justice oftentimes requires one to "speak truth to power."

The Jain soteriology, with its devaluation of the material world in the pursuit of a pure spirituality, is in many ways not conducive to the development of an environmental ethic. But the Jains also have a rich history of daily practices and attitudes that foster a much more positive engagement with the material world. Such habitual activities in relationship to the environment oftentimes underlie and inform an environmental ethic, more so than abstract moral rules and injunctions.[61]

Jains understand the wide variety of life, from single-sensed life-forms through five-sensed humans through perfected and liberated God-like souls, to form an interdependent continuity of existence. There is a moral hierarchy of life-forms, depending on the number of senses and the ability to reason. At the same time all souls, whether they inhabit single-sensed or five-sensed bodies, are in their essential natures ontologically identical, and so there is a denial that this hierarchy has any ultimate value. This combination of a context-sensitive ethic of differentiated bodies and abilities, with a universal ethic of the potential of each soul, leads to a distinctively Jain understanding of the relation between the "human" and "nonhuman" realms.[62]

Jain morality is also grounded in the understanding of *karma* as tying all life-forms together in an intercausal web. Jains are therefore expected to pay attention to the ways they both positively and negatively affect all other life-forms in thoughts, words, and actions, and also through the three modalities of action, commission, and approval or disapproval.

All of this will lead to a distinctive Jain environmental ethic. Such an ethic, however, will have to come to terms with the social location of the Jains, many of whom are among the socioeconomic elite of India (and increasingly in Britain and North America as well). It will also have to address their geographical location as a community that has traditionally emphasized a willingness to move, either in pursuit of economic opportunities or to prevent monastic attachment to any single place. Some of these economic, social, and mendicant traditions may need to be reevaluated in light of the growing recognition that solutions to ecological problems frequently require a long-term commitment to a particular place and community.

Some readers might argue that this set of notes and queries is idiosyncratic, more reflective of those questions which this particular author finds challenging than an objective assessment of the full range of environmental thought. I would contend, on the contrary, that each of the questions raises an important issue for environmentally engaged Jains to consider carefully. Some will be more productive of Jain answers than others. Some call for answers in terms of Jain practice, others in terms of reflection on the assumptions and implications of the Jain worldview. All will aid in the development of a self-conscious and therefore potentially effective Jain environmental ethic.

Notes

I thank Alan Babb, Surendra Bothara, Chris Chapple, Paul Dundas, and Dave Woodyard for their comments and insights, which have helped improve this essay. I further thank all the students who have taken a course entitled "Religion and Nature" that I have offered at Harvard and Denison. I would also like to thank those in attendance at the Harvard conference on Jainism and ecology, as well as the many other Jains in India and North America with whom I have discussed these and related issues. Their questions and analyses have been invaluable in the development of my thoughts on these issues

1. Wendell Berry, *Home Economics* (San Francisco: North Point Press), 6–20.

2. Ibid., 20.

3. Ibid.

4. In line with most recent scholarship, in this essay I use the spelling "Jain" rather than "Jaina," as it reflects the spelling and pronunciation favored by most Jains themselves.

5. I have not encountered the expression "engaged Jainism" yet among Jain friends. I import the concept from "engaged Buddhism," on which see *The Path of Compassion: Writings on Socially Engaged Buddhism*, ed. Fred Eppsteiner, 2d ed. (Berkeley: Parallax Press, 1988); Kenneth Kraft, "Prospects of a Socially Engaged Buddhism," in *Inner Peace, World Peace: Essays on Buddhism and Nonviolence*, ed. Kenneth Kraft (Albany: State University of New York Press, 1992), 11–30; *Engaged Buddhism: Buddhist Liberation Movements in Asia*, ed. Christopher S. Queen and Sallie B. King (Albany: State University of New York Press, 1996); and *Engaged Buddhism in the West*, ed. Christopher S. Queen (Boston: Wisdom Publications, 2000). See also the quarterly journal *Turning Wheel*, published by the Buddhist Peace Fellowship, as well as its home page, www.bpf.org.

6. The proceedings of the conference will be published in the journal of the Department of Philosophy of the University of Rajasthan.

7. There is an extensive vernacular Jain discourse on environmentalism, most of which in an uncritical and apologetic tone argues that Jainism is and long has been an environmental tradition. This discourse is very similar to the English discourse Anne Vallely discusses in her chapter in this book; it is not clear to me, however, whether these discourses have arisen independently or not. The vernacular discussion of Jainism as an environmental religion is found in the popular literature of all four Jain traditions. For example, a special issue of the Digambara quarterly *Prākṛt Vidyā* (volume 3, numbers 1–3, April-December 1991), published from Udaipur, on environmental harmony and vegetarianism (*paryāvaraṇ-saṅtulan evaṃ śākāhār*) contains thirteen articles, such as Śubhū Paṭvā, "What Is Environmental Culture?" ("Kyā hai Paryāvaraṇ Saṃskṛti?"); Prem Suman Jain, "Environment: The Religious Basis of Protection" ("Paryāvaraṇ: Saṃrakṣaṇ kā Dhārmik Ādhār"); Rājmal Loṛhā, "Environmental Pollution: Who is Responsible?" ("Paryāvaraṇ-Pradūṣaṇ: Kaun Jimmevār?"); Surendra Botharā, "Ācārāṅga: The Oldest Environmental Text" ("Ācārāṅg: Paryāvaraṇ kā Prācīntam Granth"); and Lakṣmīcand Saroj, "The Life of Environmentalism Is Vegetarianism" ("Paryāvaraṇ kā Prāṇ Śākāhār"). The May 1992 issue of the Digambara monthly *Tīrthaṅkar* (vol. 22, no. 1), published from Indore, contained

several articles on environmental issues, most notably Surendra Botharā, "The Environment and Jain Responsibility" ("Paryāvaraṇ aur Jain Dāyitva"). The Sthānakvāsī monthly *Jinvāṇī*, published from Jaipur, has regularly carried articles on the environment during the past decade, such as Prem Suman Jain, "The Role of Jainism in National Culture and Environmental Protection" ("Rāṣṭrīya Saṃskṛti evaṃ Paryāvaraṇ-Samrakṣaṇ mẽ Jain Dharm kī Bhṇmikā"), vol. 44, no. 2 (February 1987): 58–60; Gulābsiṃh Darṛā, "Protection of the Environment Is Protection of Life" ("Paryāvaraṇ kī Rakṣā: Jīvan kī Rakṣā"), vol. 55, no. 11 (November 1994): 32–34; and Sūrajmal Jain, "Environmental Protection in Jain Philosophy" ("Jain Darśan mẽ Paryāvaraṇ Saṃrakṣaṇ"), vol. 55, nos. 5–7 (May–July, 1998): 45–48, 19–22, 21–30. See also Bhagcandra Jain, *Jainism and the Environment* (*Jain Dharm aur Paryāvaraṇ*) (Delhi: New Bharatiya Book Company, 2001).

8. This replicates a pattern that has existed for too long in most aspects of Jain studies; see Padmanabh S. Jaini, "The Jainas and the Western Scholar," *Sambodhi* 5, nos. 2–3 (July–October 1976): 121–31, reprinted in his *Collected Papers on Jaina Studies* (Delhi: Motilal Banarsidass, 2000), 23–36.

9. The phrase is that of L. M. Singhvi, *The Jain Declaration of Nature* (reprint, Cincinnati: Federation of Jain Associations in North America, 1990), 5 (reprinted as the appendix to this volume). Similar statements can be found in the writings of Michael Tobias (*Life Force: The World of Jainism* [Berkeley: Asian Humanities Press, 1991]; *A Vision of Nature: Traces of the Original World* [Kent, Ohio: Kent State University Press, 1995]), as well as in many Jain publications aimed at English-speaking Jains in India, Europe, and North America.

10. Harold Coward, "New Theology on Population, Consumption, and Ecology," *Journal of the American Academy of Religion* 65 (1997): 261. In addition to Coward's article, my thinking here has been shaped by the comments of Ashok Aklujkar in a different yet analogous context, that of a workshop on Vedanta and conflict resolution held by the Dharam Hinduja Indic Research Center at Columbia University in October 1994.

11. I am here stating a proposition concerning the relationships among language, culture, and thought that is not universally accepted. The debates concerning these relationships in terms of "human rights" and "justice" in the contexts of Buddhism and Confucianism illustrate the basic positions of whether or not a concept can exist in a culture if there is no exact word or word cluster for it. See, in the case of Buddhism, Damien V. Keown, "Are There Human Rights in Buddhism?" and Craig K. Ihara, "Why There Are No Rights in Buddhism: A Reply to Damien Keown," both in *Buddhism and Human Rights*, ed. Damien V. Keown, Charles S. Prebish, and Wayne R. Husted (Richmond, Va.: Curzon Press, 1998), 15–41 and 43–51; and, in the case of Confucianism, Henry Rosemont, Jr., "Why Take Rights Seriously? A Confucian Critique," and Roger T. Ames, "Rites as Rights: The Confucian Alternative," both in *Human Rights and the World's Religions*, ed. Leroy S. Rouner (Notre Dame: University of Notre Dame Press, 1988), 167–82 and 199–216; Henry Rosemont, Jr., "Rights-Bearing Individuals and Role-Bearing Persons," in *Rules, Rituals, and Responsibilities: Essays Dedicated to Herbert Fingarette*, ed. Mary I. Bockover (La Salle, Ill.: Open Court, 1991), 71–101; and Sumner B. Twiss, "A Constructive Framework for Discussing Confucianism and Human Rights," in *Confucianism and Human*

Rights, ed. Wm. Theodore de Bary and Tu Wei-Ming (New York: Columbia University Press, 1998), 27–53.

12. Gary Snyder, *The Practice of the Wild* (San Francisco: North Point Press, 1990), 8–9.

13. For some illustrative examples of the fast-expanding literature of environmental history, see, among many others, Lawrence Buell, *The Environmental Imagination: Thoreau, Nature Writing, and the Formation of American Culture* (Cambridge, Mass.: Harvard University Press, 1995); William Cronon, *Changes in the Land: Indians, Colonists, and the Ecology of New England* (New York: Hill and Wang, 1983), *Nature's Metropolis: Chicago and the Great West* (New York: W. W. Norton, 1991), and *Uncommon Ground: Toward Reinventing Nature*, ed. William Cronon (New York: W. W. Norton, 1995); J. R. McNeill, *Something New under the Sun: An Environmental History of the Twentieth-Century World* (New York: W. W. Norton, 2000); Carolyn Merchant, *The Death of Nature: Women, Ecology, and the Scientific Revolution* (San Francisco: Harper and Row, 1980), and *Ecological Revolutions: Nature, Gender, and Science in New England* (Chapel Hill: University of North Carolina Press, 1989); Roderick Nash, *Wilderness and the American Mind*, 3d ed. (New Haven: Yale University Press, 1982); Barbara Novak, *Nature and Culture: American Landscape and Painting 1825–1875* (New York: Oxford University Press, 1980); Max Oelschlaeger, *The Idea of Wilderness: From Prehistory to the Age of Ecology* (New Haven: Yale University Press, 1991); and Simon Schama, *Landscape and Memory* (New York: Vintage, 1996). The work of William R. LaFleur ("Saigyō and the Buddhist Value of Nature," *History of Religions* 13 [1973]: 227–48) and Miranda Shaw ("Nature in Dōgen's Philosophy and Poetry," *Journal of the International Association of Buddhist Studies* 8 [1985]: 111–32) on Japan also provide models for such scholarship.

Environmental history is not as well-established a field in South Asia. For an overview of the literature, see the excellent introductions to *Nature, Culture, Imperialism: Essays on the Environmental History of South Asia*, ed. David Arnold and Ramachandra Guha (Oxford: Oxford University Press, 1995) and *Nature and the Orient: The Environmental History of South and Southeast Asia*, ed. Richard H. Grove, Vinita Damodaran, and Satpal Sangwan (Delhi: Oxford University Press, 1998). Both of these works, as well as Madhav Gadgil and Ramachandra Guha, *This Fissured Land: An Ecological History of India* (Delhi: Oxford University Press, 1992), focus much more on material and political-economic analyses than cultural analyses. Foundations for a cultural history of nature and the environment in South Asia can be found in works such as Ananda K. Coomaraswamy, *The Transformation of Nature in Art* (Cambridge, Mass.: Harvard University Press, 1934); Alan Entwistle, "The Cult of Krishna-Gopāl as a Version of Pastoral," in *Devotion Divine: Bhakti Traditions from the Regions of India. Studies in Honour of Charlotte Vaudeville*, ed. Diana L. Eck and Françoise Mallison (Groningen: Egbert Forsten; and Paris: École Française d'Extrême-Orient, 1991), 73–90; Peter Gaeffke, "The Concept of Nature in the Literature of Bengali, Hindi, and Urdu around 1900," *Studien zur Indologie und Iranistik* 10 (1985): 79–101; A. K. Ramanujan, *Poems of Love and War* (New York: Columbia University Press, 1985); David Sopher, "Place and Landscape in Indian Tradition," *Landscape* 29 (1986): 2–9; and Francis Zimmermann, *The Jungle and the Aroma of*

Meats: An Ecological Theme in Hindu Medicine, trans. Janet Lloyd (Berkeley: University of California Press, 1987).

14. See John E. Cort, "Recent Fieldwork Studies of the Contemporary Jains," *Religious Studies Review* 23 (1997): 103–11, and Paul Dundas, "Recent Research on Jainism," *Religious Studies Review* 23 (1997): 113–19, for bibliographic introductions into this scholarship.

15. "Jains" and "Jainism" as unmarked, nonsectarian categories have emerged as social realities only in recent decades in North America and Europe, as the Jain community has actively striven to eschew sectarian differences. In India, however, whenever one uses the words "Jain" and "Jainism," there always is (or should be) a hidden sectarian qualifier.

16. See the similar comments by Lance E. Nelson, "Reading the *Bhagavadgītā* from an Ecological Perspective," in *Hinduism and Ecology: The Intersection of Earth, Sky, and Water*, ed. Christopher Key Chapple and Mary Evelyn Tucker (Cambridge, Mass.: Center for the Study of World Religions, Harvard Divinity School, 2000), 135, 157–58 n. 31.

17. On the vows, see Paul Dundas, *The Jains* (London: Routledge, 1992), 135–38.

18. On *pratikramaṇa*, see John E. Cort, *Jains in the World: Religious Values and Ideology in India* (New York and Delhi: Oxford University Press, 2001), 123–24; and James Laidlaw, *Riches and Renunciation: Religion, Economy and Society among the Jains* (Oxford: Clarendon Press, 1995), 190–215. On the eighteen *pāpasthāna*s, or most serious transgressions, see Laidlaw, *Riches and Renunciation*, 211–12; and R. Williams, *Jaina Yoga* (Oxford: Oxford University Press, 1963), 206–7.

19. The clearest example of such programmatic recovery of lost or neglected voices has been the extensive work of feminist scholarship over recent decades. For examples of such efforts within environmental history, see Lynn White, Jr.'s advocacy of Francis of Assisi ("The Historical Roots of Our Ecologic Crisis," *Science* 155 [1967]: 1203–7), and Ramachandra Guha's and Juan Martinez-Alier's advocacy of M. K. Gandhi, Nicholas Georgescu-Roegen, and Lewis Mumford as precursors of modern environmental thought (*Varieties of Environmentalism: Essays North and South* [London: Earthscan, 1997], 153–201).

20. Gary Paul Nabhan, *Cultures of Habitat: On Nature, Culture, and Story* (Washington, D.C.: Counterpoint, 1997), 319.

21. My discussion here summarizes what is argued at much greater length in my *Jains in the World*.

22. Jainism is by no means unique among the world's religious traditions in this ideological focus on a soteriology that in the end either devalues or negates altogether the material world of day-to-day existence. See the similar observations concerning Advaita and other orthodox Brāhmaṇical metaphysics by Vasudha Narayanan, "'One Tree Is Equal to Ten Sons': Some Hindu Responses to the Problems of Ecology, Population, and Consumerism," in *Visions of a New Earth: Religious Perspectives on Population, Consumption, and Ecology*, ed. Harold Coward and Daniel C. Maguire (Albany: State University of New York Press, 2000), 111–29; Lance E. Nelson, "The Dualism of Non-Dualism: Advaita Vedānta and the Irrelevance of Nature," in *Purifying the Earthly Body of God: Religion and Ecology in Hindu India*, ed. Lance E. Nelson (Albany: State University of New York Press, 1998), 61–88; Nelson, "Read-

ing the *Bhagavadgītā* from an Ecological Perspective"; and Arvind Sharma, "Attitudes to Nature in the Early Upaniṣads," in *Purifying the Earthly Body of God: Religion and Ecology in Hindu India*, ed. Lance E. Nelson (Albany: State University of New York Press, 1998), 51–60.

23. See, for examples, Marcus Banks, "Representing the Bodies of the Jains," in *Rethinking Visual Anthropology*, ed. Marcus Banks and Howard Morphy (New Haven: Yale University Press), 216–39; Padmanabh S. Jaini, "Fear of Food? Jain Attitudes on Eating," in *Jain Studies in Honour of Jozef Deleu*, ed. Rudy Smet and Kenji Watanabe (Tokyo: Hon-no-Tomosha), 339–53, reprinted in *Collected Papers on Jaina Studies* (Delhi: Motilal Banarsidass, 2000), 281–96; and Laidlaw, *Riches and Renunciation*, 151–286.

24. For the sake of readers who are not familiar with Jainism, I here summarize these two realms of value (adapted from my *Jains in the World*, 6–7):

> The *mokṣa-mārga* is the orthodox ideology of the path to liberation, as symbolized in the temple image of the liberated Jina, who in the past traveled the path to liberation and then taught it to the world, and the living figure of the world-renouncing ascetic mendicant. It consists of correct faith in the truth and efficacy of the Jain teachings, correct knowledge and understanding of those teachings, and correct conduct that leads one along the path to liberation. These are the three jewels that make up the Jain religion, to which are often added a fourth jewel of correct asceticism, which emphasizes the central place of ascetic practices in Jain soteriology. Traveling the *mokṣa-mārga* involves increased faith in and knowledge of the true nature of the universe. It also involves conduct designed simultaneously to reduce and ultimately halt the influx of karma and to scrub away the accumulated karma of this and previous lives. Karma is the "problem" according to the *mokṣa-mārga* ideology. According to the Jains karma is a subtle material substance. Every unenlightened thought, deed and word causes karma to stick to the soul like invisible glue. This karma both creates ignorance of the true nature of the universe and blocks the inherent perfection of the soul. The soul in its pure, unfettered state is characterized by the four infinitudes of knowledge (*jñāna*), perception or intuition (*darśana*), bliss (*sukha*), and power (*vīrya*). When finally freed of its enslaving karma, the enlightened and liberated soul floats to the top of the universe to exist forever as a self-sufficient monad absorbed in the four infinitudes. The *mokṣa-mārga* thus necessitates the increased isolation of the soul, and emphasizes separation of the individual from worldly ties and interactions.
>
> The realm of wellbeing is not ideologically defined, and is therefore somewhat more difficult to delineate. . . . Whereas the *mokṣa-mārga* involves the increasing removal of oneself from all materiality in an effort to realize one's purely spiritual essence, wellbeing is very much a matter of one's material embodiment. It is marked by health, wealth, mental peace, emotional contentment, and satisfaction in one's worldly endeavors. . . . The "goal" of this realm, to the extent that it is at all goal-oriented, is a state of harmony with and satisfaction in the world, a state in which one's social, moral, and spiritual interactions and responsibilities are properly balanced.

25. See also the chapter by Paul Dundas in this volume.

26. On the context-sensitivity of South Asian ethics, see Charles Hallisey, "Ethical Particularism in Theravāda Buddhism," *Journal of Buddhist Ethics* 3 (1996): 32–43; and A. K. Ramanujan, "Is There an Indian Way of Thinking? An Informal Essay," *Contributions to Indian Sociology*, n.s., 23 (1989): 41–58, reprinted in *The Collected Essays of A. K. Ramanujan*, ed. Vinay Dharwadker (Delhi: Oxford University Press, 1999), 34–51.

27. Here I think of the many problems unnecessarily created by Jains who have tried to establish the "scientific bases" of Jain doctrine and worldview, but who have worked with an inadequate understanding of the presuppositions and operating principles of the episteme of the scientific method.

28. An excellent overview of these various streams within environmental thought and practice is Carolyn Merchant, *Radical Ecology: The Search for a Livable World* (New York: Routledge, 1992). The reader should be aware that in my discussion I have played down the many and oftentimes strident disagreements among different environmental thinkers. In my notes I have indicated literature to which the interested reader can go for more on the background and expression of the various questions raised by the various environmentalist voices. I have tried to provide foundational works in these subfields, as well as examples of the most recent scholarship, which will also provide the most recent bibliographies. This is by no means a comprehensive bibliography, but rather a starting point for further reading.

29. Dundas, *The Jains*, 80–83; and Jaini, *The Jaina Path of Purification*, 97–106.

30. On the Jain biology of *kāya*s, see Jaini, *The Jaina Path of Purification*, 108–11.

31. See Bill Devall and George Sessions, *Deep Ecology: Living as if Nature Mattered* (Salt Lake City: Peregrine Smith Books, 1985); *The Deep Ecology Movement: An Introductory Anthology*, ed. Alan Drengson and Yuichi Inoue (Berkeley: North Atlantic Books, 1995); *Beneath the Surface: Critical Essays in the Philosophy of Deep Ecology*, ed. Eric Katz, Andrew Light, and David Rothenberg (Cambridge, Mass.: MIT Press, 2000); Merchant, *Radical Ecology*, 85–109; and *Deep Ecology for the Twenty-first Century*, ed. George Sessions (Boston: Shambhala, 1995).

32. The issue of whether diversity has intrinsic or relative value is, of course, not limited to the relationships between humans and nonhumans. Ashis Nandy has recently raised it as the central moral issue in India's social and cultural future as well: Ashis Nandy, "Reimagining India's Present," *Manushi* 123 (March-April 2001): 17–21.

33. See Merchant, *Radical Ecology*, 61–82; Roderick Nash, *The Rights of Nature: A History of Environmental Ethics* (Madison: University of Wisconsin Press, 1989); Christopher D. Stone, *Earth and Other Ethics: The Case for Moral Pluralism* (New York: Harper and Row, 1987).

34. Marie-Claude Mahias, *Délivrance et convivialité: Le système culinaire des Jaina* (Paris: Éditions de la Maison des Sciences de l'Homme, 1985).

35. Jain biology distinguishes between plant bodies that contain each a single soul (*pratyeka*) and those that contain aggregated souls (*sādhāraṇa*).

36. Within the Jain ontology of four possible realms of birth (*gati*), plants and animals (*tiryañca*) are collectively considered as one realm; the other three are humans (*manuṣya*), heavenly beings (*deva*), and infernal beings (*nāraki*).

37. Translated from text in *Śrāddha-Pratikramaṇa Sūtra (Prabodha Ṭīkā)*, ed. Paṅnyās Bhadraṅkarvijaygaṇi and Muni Kalyāṇprabhvijay (Bombay: Jain Sāhitya Vikās Maṇḍal, 1977), 2:120–21.

38. Here I am thinking of Arthur Waskow's (*Seasons of Our Joy* [Boston: Beacon Press, 1982]) creative reading of Jewish rituals from a perspective concerned with social justice, and of the adaptations of traditional precepts by engaged Buddhists, such as Robert Aitken (*The Mind of Clover: Essays in Zen Buddhist Ethics* [San Francisco: North Point Press, 1984], 3–104), Bernie Glassman (*Bearing Witness: A Zen Master's Lessons in Making Peace* [New York: Bell Tower, 1998], 214–17), and Thich Nhat Hanh (*Interbeing: Fourteen Guidelines for Engaged Buddhism*, ed. Fred Eppsteiner, rev. ed. [Berkeley: Parallax Press, 1993]), to encompass also social justice, as models for a "green *pratikramaṇa*."

39. See Cort, *Jains in the World*, chap. 3.

40. The classic statement of this within a Christian ecological context is Sallie McFague, *Models of God: Theology for an Ecological, Nuclear Age* (Philadelphia: Fortress, 1987). See also her *Body of God: An Ecological Theology* (Minneapolis: Fortress, 1993).

41. This aspect of Jain cosmology might well bear fruitful comparison with other cosmologies, such as the Chinese "continuity of being" (Tu Weiming, "The Continuity of Being: Chinese Visions of Nature," in *Confucianism and Ecology: The Interrelation of Heaven, Earth, and Humans*, ed. Mary Evelyn Tucker and John Berthrong [Cambridge, Mass.: Center for the Study of World Religions, Harvard Divinity School, 1998], 105–21), the Christian "chain of being" (Arthur O. Lovejoy, *The Great Chain of Being: A Study of the History of an Idea* [Cambridge, Mass.: Harvard University Press, 1936]), and, in a different context, Wendell Berry's concept of the "great economy" (*Home Economics*, 54–75).

42. From among the vast array of ecofeminist writings, see Susan Griffin, *Woman and Nature: The Roaring Inside Her* (New York: Harper and Row, 1978); Merchant, *Radical Ecology*, 183–210; Karen J. Warren, *Ecofeminist Philosophy: A Western Perspective on What It Is and Why It Matters* (Lanham, Md.: Rowman and Littlefield, 2000); and the following volumes edited by Karen J. Warren: *Ecological Feminism* (London: Routledge, 1994), *Ecological Feminist Philosophies* (Bloomington: Indiana University Press, 1996), and *Ecofeminism: Women, Culture, Nature* (Bloomington: Indiana University Press, 1997). For Hindu expressions of ecofeminism, see Lina Gupta, "Ganga: Purity, Pollution, and Hinduism," in *Ecofeminism and the Sacred*, ed. Carol J. Adams (New York: Continuum), 99–116; Maria Mies and Vandana Shiva, *Ecofeminism* (New Delhi: Kali for Women, 1993); Rita DasGupta Sherma, "Sacred Immanence: Reflections of Ecofeminism in Hindu Tantra," in *Purifying the Earthly Body of God: Religion and Ecology in Hindu India*, ed. Lance E. Nelson (Albany: State University of New York Press, 1998), 89–131; and Vandana Shiva, *Staying Alive: Women, Ecology and Development* (London: Zed Books, 1989).

43. See Lina Gupta, "Kali, the Savior," in *After Patriarchy: Feminist Transformations of the World Religions*, ed. Paula M. Cooey, William R. Eakin, and Jay B. McDaniel (Maryknoll: Orbis Books, 1991), 15–38; Vijaya Nagarajan, "The Earth as Goddess Bhu Devi: Toward a Theory of 'Embedded Ecologies' in Folk Hinduism," in *Purifying the Earthly Body of God: Religion and Ecology in Hindu India*, ed. Lance

E. Nelson (Albany: State University of New York Press, 1998), 269–95; and Sherma, "Sacred Immanence."

44. The classic ecofeminist expression of these connections is Susan Griffin, *Pornography and Silence: Culture's Revenge against Nature* (New York: Harper and Row, 1981).

45. A start to this large project has been made in works such as Sherry Elizabeth Fohr, "Gender and Chastity: Female Jain Renouncers" (Ph.D. diss., University of Virginia, 2001); Savitri Holmstrom, "Towards a Politics of Renunciation: Jain Women and Asceticism in Rajasthan" (M.A. diss., University of Edinburgh, 1998); Padmanabh S. Jaini, *Gender and Salvation: Jaina Debates on the Spiritual Liberation of Women* (Berkeley: University of California Press, 1991); M. Whitney Kelting, *Singing to the Jinas: Jain Laywomen, Maṇḍaḷ Singing, and the Negotiations of Jain Devotion* (New York: Oxford University Press, 2001); Josephine Reynell, "Prestige, Honour and the Family: Laywomen's Religiosity amongst the Śvetāmbar Mūrtipūjak Jains in Jaipur," *Bulletin d'Études Indiennes* 5 (1987): 313–59; Josephine Reynell, "Women and the Reproduction of the Jain Community," in *The Assembly of Listeners: Jains in Society*, ed. Michael Carrithers and Caroline Humphrey (Cambridge: Cambridge University Press, 1991), 41–65; N. Shântâ, *La Voie Jaina: Histoire, spiritualité, vie des ascètes pèlerines de l'Inde* (Paris: O.E.I.L., 1985); Anne Vallely, *Guardians of the Transcendent: An Ethnography of a Jain Ascetic Community* (Toronto: University of Toronto Press, 2002); and Leonard Zwilling and Michael J. Sweet, "'Like a City Ablaze': The Third Sex and the Creation of Sexuality in Jain Religious Literature," *Journal of the History of Sexuality* 6 (1996): 359–84. The study of Jain understandings of both women's religiosity and gender construction in Jainism will be further enhanced by the dissertations-in-progress of Tinamarie Jones (McMaster University) and Madhurina Shah (University of Maryland).

46. Jaini, *Gender and Salvation*.

47. Kelting, *Singing to the Jinas*.

48. Again, the literature here is vast. See Murray Bookchin, *Remaking Society: Pathways to a Green Future* (Boston: South End Press, 1990); Robert D. Bullard, *Dumping in Dixie* (Boulder: Westview Press, 1994); *Unequal Protection: Environmental Justice and Communities of Color*, ed. Robert D. Bullard (San Francisco: Sierra Club Books, 1994); Luke W. Cole and Sheila R. Foster, *From the Ground Up: Environmental Racism and the Rise of the Environmental Justice Movement* (New York: New York University Press, 2001); Ramachandra Guha and Juan Martinez-Alier, *Varieties of Environmentalism: Essays North and South* (London: Earthscan, 1997); and *Global Ethics and Environment*, ed. Nicholas Low (London: Routledge, 1999). One should note that both social ecology and Southern environmentalism have roots in various forms of Marxist and materialist analyses, which pose other questions for any Jain environmental ethic.

49. Guha and Martinez-Allier, *Varieties of Environmentalism*, 91.

50. Ramachandra Guha, "Radical American Environmentalism and Wilderness Preservation: A Third World Critique," in Guha and Martinez-Allier, *Varieties of Environmentalism*, 101.

51. Bullard, *Dumping in Dixie*.

52. Madhav Gadgil and Ramachandra Guha, *Ecology and Equity: The Use and Abuse of Nature in Contemporary India* (London: Routledge, 1995).

53. Information on Shatrunjay comes from the short article "Greening of Palitana" in *Ahimsa* 7, no. 4 (June-December 1997): 7, and from conversations with Jain friends in Gujarat. See also the essay by the environmental activist John Seed ("Spirit of the Earth: A Battle-Weary Rainforest Activist Journeys to India to Renew His Soul," *Yoga Journal* 138 [1998]: 69–71, 132–36) about similar problems in a program to reforest the sacred Hindu mountain of Arunachala in South India, in which the Australian author betrays an attitude of arrogance in his portrayal of the indifference of the Indian pilgrims to his ecological agenda.

54. For thorough expressions of this position from a Christian perspective, see Herman E. Daly and John B. Cobb, Jr., *For the Common Good: Redirecting the Economy toward Community, the Environment, and a Sustainable Future* (Boston: Beacon Press, 1989), and Paul G. King and David O. Woodyard, *Liberating Nature: Theology and Economics in a New Order* (Cleveland: Pilgrim Press, 1999).

55. The twelve lay vows are almost never taken in a formal sense by laity, and themselves are part of the *mokṣa-mārga* ideological discourse. But they are expressive of values widely understood within the Jain community.

56. Atul K. Shah, "Imprisoned by Success," *Young Jains International Newsletter* 11, no. 3 (1997): 5–6.

57. See Sukhlal Sanghavi, *Pacifism and Jainism* (Banaras: Jain Cultural Research Society, 1950). Few Jains with whom I spoke in India in May 1998 saw the Jain understanding of *ahiṃsā* as calling for public opposition to either the tests of nuclear weapons at Pokharan or the broader militarization of Indian society, although the Jaipur Hindi press did carry reports of two Terāpanthī monks speaking out against the tests at a public meeting on May 24.

58. In addition to Merchant, *Radical Ecology*, 217–22, see the following: *Home! A Bioregional Reader*, ed. Van Andruss, Christopher Plant, Judith Plant, and Eleanor Wright (Philadelphia: New Society Press, 1990); *Bioregional Assessments: Science at the Crossroads of Management and Policy*, ed. K. Norman Johnson et al. (Washington, D.C.: Island Press, 1998); and Kirkpatrick Sale, *Dwellers in the Land: The Bioregional Vision* (San Francisco: Sierra Club), 1985. See also Christian-based expressions of stewardship, such as those of Wendell Berry (*The Unsettling of America: Culture and Agriculture* [New York: Avon, 1977]) and Wes Jackson (*Becoming Native to This Place* [Lexington: University Press of Kentucky, 1994]), which share much with bioregionalism.

59. See Merchant, *Radical Ecology*, 219.

60. Jaini, *The Jaina Path of Purification*, 247–48.

61. My thoughts here have been informed by Maria Hibbets, "The Ethics of the Gift: A Study of Medieval South Asian Discourses on *Dana*" (Ph.D. diss., Harvard University, 1999). See also Pierre Bourdieu, *Outline of a Theory of Practice*, trans. Richard Nice (Cambridge: Cambridge University Press, 1977).

62. I am here simplifying a rather complex topic, for Jains have not spoken with a single voice concerning the soteriological capabilities of all souls; on this, see Padmanabh S. Jaini, "*Bhavyatva* and *Abhavyatva*: A Jain Doctrine of 'Predestination,'" in *Mahavira and His Teachings*, ed. A. N. Upadhye et al. (Bombay: Bhagavan Mahāvīra 2500th Nirvāṇa Mahotsava Samiti, 1977), 95–111; reprinted in *Collected Papers on Jaina Studies*, 95–109.

The Limits of a Jain Environmental Ethic

PAUL DUNDAS

At first blush, there would appear to be a remarkable fit between environmentalism and Jainism's teachings about the ubiquity of embodied souls in the natural world and the desirability of adopting a nonviolent stance toward them. What more striking example could there be, after all, of the promotion of a respectful attitude toward nature than Jainism's institutionalization of the ascetic's guarded and controlled mode of walking, designed to minimize the possibility of the slightest form of violence to minute life-forms, into a formula at the very center of the Jain confessional observance, supposedly binding upon all?[1] This is undoubtedly the view of *The Jain Declaration on Nature* of 1992, for whose author "the Jain ecological philosophy is virtually synonymous with the principle of *ahiṃsā* [nonviolence],"[2] and it was equally the deeply held conviction of by far the majority of the speakers at the Center for the Study of World Religion's conference of 1998 whose proceedings are recorded in this volume. Yet, there may be some grounds for suspecting that the recent equation of Jainism and environmentalism is not always so comfortable as its advocates would suggest.

Contrary to general belief, Indian concepts of nature do not necessarily derive from a monistic metaphysics. As far as Jainism is concerned, there can be no possible hypostatization of the natural world into an undifferentiated "Nature" envisaged as the source or matrix of all entities. The Jains are in actuality both metaphysical dualists and, from an ontological perspective, pluralist realists of a most radical type, with nondual monism generally being associated by the tradition's teachers with idealism.[3] Acceptance of a plurality of real entities would, of course, appear a particularly appropriate ideology upon which to ground an ecological ethic. However, the judgment of an important Jain scripture that plants, while sentient, of necessity fall into the negative soteriological category of "having false belief" (*mithyādṛṣṭi*) is a significant pointer to the broad Jain perspective on the natural world.[4]

Nature and its manifold qualities, seen and unseen, have no autonomous value for Jainism, but instead are linked to the various gradations of Jain epistemology, in that the more spiritually advanced the individual, the more developed his knowledge, and consequently the greater awareness he possesses of the infinite and increasingly minute constituent elements of the universe. It is in fact only the fully enlightened beings, the *kevalin*s, who have obtained the perfect epistemological state of omniscience, who can fully and directly comprehend things in their totality.[5] Knowledge of the reality of nature is in these terms not an end in itself but rather a vital index of progress on a path of spiritual development that can only be followed to its full conclusion by human beings. In this respect, priority and value are assigned to the human state, and nature *qua* flora and fauna has no meaning in its own right. Instead, it is consistently envisaged in Jain literature as representing a type of diminished or inadequate humanity, with proximity to or distance from human birth being in the last resort the most significant determinant of the value of its constituent elements. Examination of the numerous Jain stories found in the canonical and postcanonical texts involving animals engaging in pious or morally positive activities makes clear how human qualities are consistently superimposed upon nonhuman protagonists.[6]

Thus far, then, I would submit that it is not enough simply to invoke the minute codifications and taxonomies of life-forms found in some of the Jain scriptures or the religion's general advocacy of the necessity for nonviolence as evidence for some sort of ancient environmentalism. But there seems to me a still more compelling objection to this. Those who would wish to place Jainism at the center of the environmentalist enterprise and, in addition, claim moral priority for the religion on the grounds of its antiquity will have to confront the fact that Jainism has a teleology of decline and subsequent renewal built into its mythical structure, which furnishes the religion's adherents with a totalizing explanation for the configuration of past, present, and future. Jain tradition is clear that, as we enter the final stages of each particular movement of the wheel of time, it is necessary and inevitable that both humankind and the natural world socially and ecologically decay. The world will be destroyed and human beings will degenerate intellectually and culturally, to be renewed subsequently with the next motion of time.[7] The tradition is unanimous that the enlightened teachers have asserted this to be the case, irrespective of the philologically derived judgment that the specific details of the dynamics of this decline may not have been fully in place textually before the medieval period. If the Jain religion is true, inasmuch as it has been revealed by these enlightened teachers, then it must be true in this respect also, so that the destruction of the environment, whether brought about by humans or spontaneously engendered, is an essential fact of existence within the Jain universe. To reject such a mythical teleology as metaphorical and thus inappropriate to the modern world must entail the restructuring or reformulation of what are for Jainism, at least from early medieval times, essential components of the tradition's message. It may be possible to present such a reformulation as a justifiable accommodation to contemporary preoccupations, but less legitimate to argue for its time-honored authenticity.

Here, by way of developing my critique, I would like to formulate an axiom with reference to Jainism that draws attention to the paradox involved in the attempt to present that tradition as an ecological philosophy.[8] The situation can be put bluntly. If humankind is the vector for the degradation of nature and its constituent ecosystems, a basic premise of modern environmentalism, then it must also be said that in Jainism it is nature that is the vector for the degradation of humankind.

To expand. The emphasis throughout Jainism has consistently been upon the danger that nature causes man through his interaction with it and his careless propensity, ultimately dependent upon *karma*, to cause violence. Even the fully enlightened person, at least in the view of the majority of Śvetāmbara Jains, cannot engage with this world without injuring living creatures, although admittedly he does not accrue negative *karma* through his actions.[9] According to the exclusively ascetic picture of the *Bhagavatī Ārādhanā* of Śivārya (early common era), violence can be brought about to living creatures by mere possession of hair and bodily cleanliness or by the simple arising of the passions.[10] As Ācārya Mahāprajña points out in what is perhaps the single best book on Jain nonviolence, destruction of the natural world has the drastic consequence of destroying the agent of destruction.[11] It is the man who injures or occludes the innately positive qualities of his self through wittingly or unwittingly engaging in morally disfiguring activity who is truly the man of violence. Nature is in this worldview perceived not as a source of comfort or succor, but as a potentially menacing force that by its mere existence has the capacity to threaten the embodied soul at all times, manifesting itself even on its external physical body. Taming of the passions, which are the cause of injury to the soul and to living creatures, through disengagement from nature is thus the sole means of gaining the ultimate goal of deliverance.

I would suggest, then, that a degree of measured reflection is required on the part of those who wish on the basis of textual sources to envisage Jainism as a primordial environmentalist religious path. Many would, of course, argue that it is fully justifiable to call upon the potential resources of a religion in order to construct a view of how that particular tradition might make a valid contribution to an issue of importance to the modern world, even if the religion in question as historically constituted may not initially appear to have much of immediate relevance to offer.[12] However, it can only be problematic to compel without adequate consideration religions and their authoritative writings, often of great antiquity and reflecting the preoccupations of now inaccessible social contexts, to conform to modern concerns. If we are truly sensitive to the complexity of the imaginative world of religious traditions, then it should go against the grain to link the parameters of a particular faith and the experience of it to its putative "values," which all too often turn out to be little more than a pared-down cluster of supposedly core teachings which in that form can be conveniently incorporated into current controversies.

I have no desire to rerun here the old scholarly debate about essentialism and its attendant problems for the study of religions, but it seems to me indisputable that to detach Jain teachings from their overall historical, practical, and mythical context in the cause of rendering them into a quasi-scientific ahistorical philosophy palatable to the modern world is completely at variance with the trajectory of informed scholarship on Asian religious traditions, including Jainism, carried out in the last fifteen years or so.[13] In making this claim, it is not my intention to pick intellectual holes in Jainism's teachings or dismay members of the Jain community currently living in the West who might wish to promote their religion's values. I do, however, feel extremely uneasy with what appears to be the continual recycling of the same quotations of decontextualized chunks gouged from the oldest scriptures (usually, in this case, the first book of the *Ācārāṅga Sūtra*) that has tended to characterize attempts to defend Jain environmentalism. Very often, what emerges from this is little more than a trivialization of Jainism into strings of platitudes.

Having given this general caveat, I would like in what follows in this chapter to consider some loosely connected textual examples from varying periods of Jain history, which I suspect may not be particularly well known. I hope to show how an awareness of emphasis and nuance, sometimes historically shifting, in areas of Jain discourse that relate to the difficulties involved in basic human activities, such as the necessity to eat, engage in business, and erect buildings, might possibly contribute to a more informed understanding of the issues at stake in projecting Jainism into the environmentalist debate.

My first example suggests that a Jain environmental ethic derived from ancient texts might well have to be uncompromising.

The Elephant Ascetics

In a chapter toward the end of the early (perhaps third century B.C.E.) canonical text the *Sūtrakṛtāṅga Sūtra*, which describes a variety of contemporary socioreligious groups to whom the Jains were opposed, there occurs a tantalizingly brief account of a form of behavior followed by individuals whom the ninth-century C.E. commentator Śīlāṅka calls the "elephant ascetics" (*hastitāpasa*).[14] These individuals are portrayed in the *Sūtrakṛtāṅga* (2.6.52) as deliberately killing one large elephant each year and living off it for the course of that year, "out of compassion for other life forms" (*sesāṇa jīvāṇa dayaṭṭhayāe*). This obscure group apparently advocated a dietary modus vivendi, which whimsically could almost be styled proto-environmentalist, in that it seems to have involved an unwillingness to overexploit the natural world from which sustenance derives, of a type which can be found in similar form among many nomadic groups to this day.[15] Śīlāṅka, however, locates this group in the context of a critique of other styles of renunciant behavior, and he presents the elephant ascetics as specifically arguing that the mode of life pursued by Jain monks was fraudulent in its supposed rejection of violence toward the natural world. The ostensible reasons for this criticism are that vegetarian ascetics such as the Jains still destroy life-forms in plants, and the mere act of mendicant wandering inevitably leads to the destruction of life. Furthermore, the elephant ascetics supposedly claimed, all forms of alms begging are subject to the cardinal fault of expectation (*āśaṃsā*), that is to say, the Jain mendicant always has some prior sense of anticipation that he will be given food.

Such attacks were no insignificant matter for the Jains since the undermining of the validity of their monastic mode of life called into question the nature of the teachings of the Jinas. The supposed doctrine or ideal of the elephant ascetics, which might not unreasonably be represented here as the protection of the greater number at the expense of the lesser in the cause of a type of ecological responsibility, is stigmatized as unsound in the *Sūtrakṛtāṅga* (2.6.53) by Ārdraka, the spokesman of the Jain position, on the grounds that the killing of even no more than one animal a year still represents the worst possible type of violence. Furthermore, it also signifies a mode of life no different from that of the householder and is thus incapable of leading to deliverance, the ultimate goal of the religion. Śīlāṅka claims that the extreme care shown by Jain monks in their movements and seeking for food (*īryāpathikī*) means that they cannot be deemed guilty of any charge of improper conduct. On the other hand, he asserts, the elephant ascetics do not simply take the life of one creature every year; in reality, their cooking of the flesh of the slaughtered elephant leads to the destruction of a large number of life-forms that both constitute and inhabit it. Such individuals do not engage in the blameless getting of food through what Śīlāṅka, following a well-known passage from the opening of the *Daśavaikālika Sūtra* (1.2–5), calls "the behavior of bees" (*mādhukaryā vṛttyā*), delicately taking nourishment at random and without violence to those who give.[16] The commentator concludes his account of the elephant ascetics by describing how homage was subsequently paid to Ārdraka and, by extension, to the Jain teaching on nonviolence, by an escaped wild elephant.

Leaving aside the accuracy of the tradition about the elephant ascetics recorded by the *Sūtrakṛtāṅga Sūtra* and Śīlāṅka, one significant point emerges at this juncture. Any alternative, non-Jain approach to the natural world would have to be regarded as inadequate if it fails to conform to the canonical Jain standard of total abstention from violence. While modern environmentalism is clearly eclectic and commendably open to a wide variety of perspectives and influences, a Jain environmentalist ethic genuinely based upon the ancient scriptures would by its nature be uncompromising in its dependence on Jain values and exclusivist through its appeal to the manifest moral superiority of Jainism and its monastic exponents.

In objection to this, it might be claimed that the famous Jain "doctrine of manypointedness" (*anekāntavāda*), usually interpreted as endowing opposing intellectual standpoints with a limited and provisional validity, could provide justification for alternative perspectives on environmentalism. However, as recent scholarship is making clear, this is to confuse the structure of *anekāntavāda* with its purpose.[17] The ultimate aim of the doctrine of manypointedness, whatever uses it may have been put to for eirenical purposes in recent years, has been to guarantee the intellectual dominance of Jainism, rather than to offer a conciliatory, quasi-tolerant sop to its opponents. There is no scope for invoking *anekāntavāda* as providing a relativistic underpinning for non-Jain versions of environmentalism or a justification for some sort of evenhandedness toward non-Jain perspectives. Jainism as intellectually constituted has a clear sense of its own superiority, and other interpretations of reality are most generally viewed in negative terms.[18]

The example of the elephant ascetics and the Jain interpretation of their behavior as an inadequate form of environmental responsibility relate to the sphere of renunciation of the world of social relations. My following two examples, however, deal with the realm of business activity from which the monk is of necessity excluded.

The Dilemmas of Business Activity

If, as environmentalists advise us, the world's ecological problems are to be linked to human's pursuit of material improvement, rather than to simple ignorance or willful malevolence, then these can in turn be clearly connected to business, commerce, and economic activity. Although, contrary to general belief, not all Jains are wealthy or, indeed, necessarily businessmen,[19] Jainism is traditionally a religion which has received patronage for a great deal of its history from members of the merchant class. Herein, as much recent scholarship has stressed, is one of the most striking features of the Jain religion: the coexistence of a renunciatory ideology followed by full-time ascetics with a wealthy lay community whose members are oriented toward the world of material well-being and, as one prosperous Jain layman in Ahmedabad once put it to me, "have business programmed in their genes." While the early Buddhist scriptures, as represented by the Pāli canon, say comparatively little about the lay state and its obligations, the Jain scriptural canon, presumably some time before its semi-closure in the fifth century C.E., came to include a good deal of material relating to lay followers. Even at this relatively early time in Jainism's history, we find an awareness of the necessity of the pious disbursement of wealth by the layman who is represented in idealized form as a figure effectively as near to monk as to householder.[20] However, we also find at this period an exemplary narrative which on the face of it appears to signal the possible dangers to the natural world of that mercantile activity which generates the layman's wealth.

The sixth "limb" (*aṅga*) of the Śvetāmbara canon, entitled "Stories of Knowledge and Righteousness" (*Jñātṛdharmakathāḥ*), is an extensive but still little-studied portion of the Jain scriptures. Chapter seventeen of this collection of stories, entitled "The Fine Horses" (*Āiṇṇe*), tells of a maritime trading expedition from the city of Hatthisīsa that is driven by a storm to a distant island called Kāliyadīva.[21] This island is full of gold, jewels, and diamonds, and is also inhabited by magnificent horses who smell the merchants when they make shore and draw back in alarm from them. However, the merchants are not on this occasion interested in the horses. Taking such of the island's riches as they can on board, along with fresh provisions, they return to Hatthisīsa. The merchants report to the king of the city about the remarkable horses they have seen on the island and he orders them to return with his household servants in order to capture them. Loading their ships with various luxurious substances and goods, they return to the island of Kāliyadīva. Unloading the cargo, the expedition employs it to build five heaps, each of which is intended to engage one of the senses. These heaps, which have ropes attached to them, are then watched from a hiding place by the ship's company.The bait makes no impression on some of the horses, who shy away from the heaps. However, others are attracted to them and are then snared by the ropes. Brought back to Hatthisīsa, these horses are broken in by the king's trainers and suffer much physical and mental torment from the binding and beating inflicted on them.

As I have recounted this scriptural story, it might seem possible to interpret it as a kind of environmentalist parable warning against the exploitation of uncorrupted nature by acquisitive economic activity and its consequent degradation by human greed. Indeed, the narrative might well be regarded as having even more point in this respect since the horses are captured because of their attraction to objects that, realistically, would only normally hold an allure for humans. Or alternatively, the story might be interpreted as pointing to the potentially corrupting influence of political authority and its worldly requirements, a theme that surfaces not infrequently in Jainism.[22] Realistically, however, creative environmentalist readings of the story must be subordinated to the interpretation furnished within the story itself by its putative narrator, the twenty-fourth Tīrthaṅkara Mahāvīra, who is in this age the channel through which the scriptures are mediated to the unenlightened.

This interpretation confirms that while the narrative of "The Fine Horses" is certainly antimaterialist, the perspective adopted within it is an exclusively ascetic one, without specific reference to the traders and king who are its protagonists. Mahāvīra points out that whoever of his monks and nuns is not attached to the objects of the senses is to be honored in this world by both ascetics and laypeople, while the conclusion pointing to the overall moral to the story, as generally occurs at the end of each of the "Stories of Knowledge and Righteousness," asserts the corollary, namely, that any Jain monk who is attached to and deluded by sense objects becomes a general object of scorn.[23]

Only very obliquely, then, can this self-interpreting story be taken as anything other than a prescription for correct ascetic behavior and withdrawal from both the social and natural worlds. It is this aspiration to withdrawal which above all characterizes Jainism as a religious path and, I would argue, also renders it an awkward tradition to link or adapt to the modern environmentalist perspective that would seek to derive both ecological and social renewal from some sort of redefined human engagement with nature. However, if the description of the trading expedition to Kāliyadīva serves as a narrative to provide allegorical confirmation of the centrality of the ascetic ideal in

Jainism, it also cannot be forgotten that, as noted above, economic activity has always been at the very basis of the Jain lay experience. Professional business activities and the morality involved therein, the subjects of the following section of this chapter, ought therefore to be a significant area in which the Jain attitude to the environment can be gauged.

Jain Business Ethics

It is a noteworthy and paradoxical feature of the many medieval Śvetāmbara Jain handbooks on idealized lay behavior that they are invariably composed by monks.[24] Business activity and the ethics to be employed by Jain businessmen are of necessity significant topics in this literature, but they tend to be located in the disparate context of normative daily lay praxis, as opposed to being viewed as independent areas of interest. A consideration of what would seem to be the one Jain manual to focus solely upon business ethics, even though it too is written by a monk, might therefore be expected to yield some interesting insights.

Ratnaśekhara Sūri's "Light on Purity of Business Activity" (*Vyavahāraśuddhiprakāśa*)[25] was composed some time in the first half of the fifteenth century, by which period the Jain community had effectively settled into that mercantile role which has largely characterized its members to this day. Although I have not carried out an extensive investigation into its sources, this work would seem to be in part derivative, with the thirteenth-century Devendra Sūri's "Daily Duty of a Layman" (*Śrāddhadinakṛtya*) providing an important source of influence. Indeed, "Light on Purity of Business Activity" opens with three verses by Devendra Sūri that clearly indicate the linkage between worldly affairs or business (*vyavahāra*; glossed as "the means of getting money and so on") and religious activity as envisaged in Jainism:

> The omniscient Jina states that purity of business activity is the basis of dharma. Purity of money in worldly existence comes about from pure business activity; through pure money there comes pure food; through pure food comes purity of body in worldly existence; with a pure body one becomes suitable for dharma. Whatever action he performs yields fruit in worldly existence.[26]

This view of religion being necessarily underpinned by economic activity and the proper use of money appearing in its ability to generate the appropriate conditions to advance on the path is reinforced by a further verse:

> There is nothing which does not succeed through economic activity [*artha*]. The intelligent should energetically master economic activity alone.[27]

Elsewhere, Ratnaśekhara quotes a verse that confirms the necessary interrelation of the external world and religion:

> Since the universe [*loka*] is the support for all who engage in dharma, so one should avoid what is at variance with the universe and what is at variance with dharma.[28]

While this statement might seem to express some sort of inchoate ecological ethic, it effectively offers little more than the truism that all human activity is located within the cosmos as described by the Jain scriptures. In fact, "Light on Purity of Business Activity" throughout does not evince any serious concern with the implications of business upon the natural world. Rather, it is the nature of moral interaction between humans and the possible dangers that misuse of money can cause in relations between individuals that are the subjects of its illustrative narratives, and we can only speculate that the promotion of ethically positive attitudes within business activity might subsequently translate into care for the environment. There is certainly no sense conveyed by Ratnaśekhara of the Jain businessman being encouraged to look beyond his own economic and karmic concerns to the care of the natural world beyond.

As is well known, Jain works on lay activity describe a group of fifteen forbidden business professions that could be regarded as fostering violence through promotion of the destruction of life-forms of many varieties. These can be listed as follows: gaining livelihood from charcoal, destroying plants, carts, transport fees, and hewing and digging; trade in animal by-products, lac and similar substances, alcohol and forbidden foodstuffs, men and animals, and destructive articles; work involving millings, mutilation, the use of fire, the use of water, and breeding and rearing animals.[29] All these activities were envisaged in negative terms because of the violence and pain likely to be inflicted on every kind of living creature, from five-sensed humans and animals to minute organisms inhabiting earth, air, water, and fire. D. N. Bhargava, in a paper in P. S. Jain and R. M. Lodha's somewhat misleadingly titled collection *Medieval Jainism: Culture and Environment*, has made a case for these occupations having been proscribed by Jain teachers because of their polluting effect on the environment.[30] However, that this ban arose purely from ecological concerns as opposed to being the logical outcome of an ascetic worldview that associates spiritual decline with unwitting interaction with the natural world is debatable.

In his treatment of the prohibited professions, both in "Light on Purity of Business Activity" and elsewhere, Ratnaśekhara is innovative in one significant respect. Rather than denouncing the fifteen professions, he refers instead, and for the first time in Śvetāmbara tradition,[31] to seven legitimate means of gaining a livelihood and attempts to accentuate their positive dimensions. Whether this represents a turning point in the manner in which Jainism envisaged the lay state is a matter to be left open, but it certainly does seem to signal a more liberal conceptualization of the layperson's necessary requirement to support himself economically. In particular, it must be noted that included within Ratnaśekhara's list alongside trade, medicine, craft, service, and (nonmonastic) begging are agriculture and animal husbandry. The two greatest of the earlier monastic legislators, Haribhadra (sixth century?) and Hemacandra (eleventh century), had both stigmatized as improper activities the keeping of animals and cultivation of the ground with a plough in such a way as to injure *pṛthivīkāyas*, the souls living in the earth.[32] Ratnaśekhara in fact does state that these two professions are "not appropriate to the intelligent" (*na vivekijanocite*).[33] However, he continues, if someone is to engage in them, he should try to employ the quality of compassion (*dayālutā*).

In the preceding example, we find an earlier Jain prescription relating to the negative consequences of digging in the earth being altered by a later Jain authority in apparent acceptance of the inevitabality of certain types of economic activity, in this case agricultural, being followed by the Jain laity. Even more inevitable for the lay community was the need to dig in the ground in order to construct wells and the foundations for buildings. This is the theme of the next section of this chapter.

The Digging of Wells and the Building of Temples

The preoccupations of the ancient Jain scriptures were obviously not always identical to those of later times, even though the earliest texts have generally been regarded as providing a kind of theoretical template for a correctly informed view of the world and action based thereon. As Jainism progressively developed into a religious culture more and more embedded in society, it became pressing to consider how certain basic human requirements might have to be reassessed in the context of the principle of nonviolence.

One of the most obvious points where the conflicting interests of environmentalism, in terms of compassion for and protection of the natural world, and the human need to survive coincide is in respect to the construction of buildings. In fact, the erection of buildings was viewed as a potential problem from comparatively early on in Jainism, as evinced by the canonical text "Question and Answers" (*Praśnavyākaraṇa*), which contains a list of structures whose erection leads to the destruction of life-forms.[34] By the medieval period, the issue is discussed not so much in terms of the basic human need for housing and shelter, although these are necessarily involved, but rather with reference to the obligation to worship the Jina embodied in the form of an image through physical means that specifically entail the erection of temples, of necessity involving the wood of chopped-down trees and the use of other "living" material, most obviously cut flowers. Both these activities are referred to as *dravyastava*, literally, "praise (of the Jinas) with material objects." Jain teachings as embodied in the early scriptures consistently emphasize that digging in the ground inevitably involves destruction of life-forms, as does the cutting of plants and the use of materials for building and worship. Consequently, temple building and image worship have been contentious issues within the Jain community for centuries. The reforming Sthānakvāsī sect (seventeenth century) and, following them, the Terāpanthīs (eighteenth century), who advocate a strong scripture-based style of Jainism, have objected strongly to the notion of *dravyastava* on the grounds of the violence entailed, although they have not, it must be said, called into question the act of building itself. However both these sects are numerically in the minority compared to "temple-worshiping" (*mandirmārgī*) Jains. So how can such violence to the environment through digging be reconciled with the necessity both to build temples and to offer worship?

Haribhadra seems to have been the first Jain writer seriously to concern himself with the issue of the violence potentially involved in building temples and worshiping therein.[35] His approach to this difficulty is to invoke an ethic of intention and so argue that building must be linked from the start to moral qualities, such as restraint. In line with an old canonical affirmation, he asserts that there cannot be violence involved in any activity relating to religion when there is genuine "striving" (*yatanā*) for cessation (*nivṛtti*) from the destruction of life-forms.[36] Furthermore, he argues, the wood that is used to build the temple does not really involve the destruction of life-forms anyway, for the simple reason that it is bought, as opposed to being cut down, by the Jain devotee.[37]

In an attempt to strengthen what might be taken by some as an unconvincing argument, Haribhadra employs two analogies which Śvetāmbara intellectuals referred to for over a millennium. The first is the "the analogy of the snake." A woman sees her infant son playing near a hole and, fearing that he might fall in, goes over to get him. At that moment a snake emerges from the hole, so the woman has to pull her son away violently. The moral is that even though the child experiences pain through being dragged away, there would have been still more pain both for him and his mother if he had been bitten by the snake. In the same way, the violence to the environment entailed in building a temple is outweighed by the virtues to be gained from worshiping the Jinas.[38]

The second analogy, that of digging a well, is referred to by Haribhadra at *Pañcāśaka* 6.42, although this is in actuality a quotation of an earlier verse found in the *Āvaśyaka Niryukti*, the collection of mnemonic verses on the *sūtra* describing the "Obligatory Actions" (*Āvaśyaka*).[39] According to this verse, *dravyastava*, or material worship, is, in the analogy of the digging of a well, appropriate for those who are not fully restrained—in other words, laypeople—in order to diminish *saṃsāra*. Haribhadra in his commentary on the *Āvaśyaka Niryukti* explains the verse as follows:

> Some people living in a house in a newly built settlement are overcome with thirst because there is not much water and so dig a well to dispel it. Although their thirst increases as they work and they become filthy

with mud and dirt, nonetheless their thirst and the dirt (ultimately) disappear because of the water which arises from their efforts, and they and other people are happy for the remaining time. So is it the same with regard to worship with material objects (*dravyastava*). Although there is non-restraint in the activity (of building the temple and cutting flowers for worship), nevertheless there comes about purity of the modification of the agent's personality (*pariṇāmaśuddhi*) by which he completely destroys any other (negative modification) which has accrued to him through lack of restraint. So worship with material objects should be performed by those who are not fully restrained on the grounds that it has a positive consequence and results in greater destruction of karmic matter.[40]

The eleventh-century commentator on the *Pañcāśaka*, Abhayadeva Sūri, explains succinctly what the analogy signifies: "Anything which brings about other more significant (positive) qualities, even though itself innately faulty, can be resorted to. Digging a well and worship with material objects (*dravyastava*) are the same in this respect."[41] In other words, a lesser evil, in this case digging in the earth, is outweighed by the greater goods of water made available to human beings and worship offered to the Jinas.

In his commentary on the *Āvaśyaka Niryukti*, Haribhadra regards the discomfort of the physical exertion carried out by laymen as the necessary evil in digging a well, rather than the destruction of life-forms. However, in a reference to the well analogy in another of his works, he makes it clear that it is the actual violence (*ārambha*) involved in thus providing water that can also be justified.[42] For Yaśovijaya (1624–1688), a great Śvetāmbara Jain monastic intellectual who discussed this same issue long after Haribhadra in a short treatise entitled "Clarifying the Analogy of the Well" (*Kūpadṛṣṭāntaviśadīkaraṇa*), it is also the violence of the act of digging, as well as the exhaustion, thirst, and being covered in mud, that is the point at issue. Like Haribhadra, Yaśovijaya invokes an ethic of intention involved in the act of digging in the ground, and he further invokes the principle of "for the greater good" and the generation of a positive disposition (*śubhabhāva*) through sound ethical action.[43] He points out that this is not an issue that has anything to do with monks, but that it relates to the householder who destroys earth creatures through engaging in agriculture and so on for the sake of family, material prosperity, and home. Digging a well or worshiping the Jinas with cut flowers may both be, on the face of it, slightly at fault, if judged in exclusively doctrinal terms, but ultimately these actions are meritorious because they bring about a morally positive disposition in the agent.[44] Most markedly, Yaśovijaya makes a comment that might well inform an environmentalist perspective for those Jains struggling to reconcile the often radically different requirements of the religious and business lives: "even in activity relating to religion, a little evil caused by violence can be approved."[45] As Yaśovijaya points out, even the venerable scripture, the *Bhagavatī*, substantiates this by describing how, despite there being an element of evil involved if a lay donor should give inappropriate food to morally upright monks, there still comes about removal of *karma* (*nirjarā*) for him through the piety informing the action.[46]

The point being made by Yaśovijaya is significant for lay Jains engaging in all sorts of basic and unavoidable activities. When there does exist broadly correct behavior (*vidhi*) represented by care in one's actions and proper motivation conjoined with true devotion, then an element of imperfect procedure in the course of some action does not invalidate them. Genuine violence only comes about when there is lack of attention to properly prescribed action.[47] In the words of a verse of Haribhadra quoted by Yaśovijaya: "it is delusion for those who (throughout their lives) destroy life-forms to support their bodies physically (i.e., by eating) also (to argue for the necessity of) not engaging in the destruction of life-forms in the course of worship of the Jinas."[48] The type of violence involved in eating, as well as in digging wells or offering worship to the Jinas, is, then, not really violence in the true sense of the action.[49]

Conclusion

What can be extrapolated from this? It should be clear that while the ancient principles of "nonviolence is the highest religious duty" (*ahiṃsā paramo dharmaḥ*) and "mutual support toward all living creatures" (*parasparopagraho jīvānām*), as delineated both in ancient texts and in the recent *Jain Declaration on Nature*, do provide the modern Jain environmentalist movement with the components of a morally unified ideological underpinning, Jainism has more than one voice and historically nuanced view on the issue of violence throughout its history.

In the ascetic worldview of the *Sūtrakṛtāṅga Sūtra*, an act of violence, in the case mentioned above as carried out by the elephant ascetics, was indefensible under any circumstances, even if it could be argued that it had a beneficial effect upon the environment through minimizing the slaughter of animals. In a much later period, that of Yaśovijaya in the seventeeth century, in which the requirements and difficulties of the Jain laity were fully acknowledged and legislated for, violence, at least to organisms in the earth, could be excused on the grounds that it ensured both physical survival through providing water (and food) and a fully Jain life through enabling temple worship of the Jinas. In other words, we find Jain tradition recording one situation in which a certain degree of positively motivated environmental exploitation is totally reprehensible, and another in which it might be useful and understandable. Witness also Ratnaśekhara's differing perspective on the issue of valid lay professional activity from that of earlier Jain teachers.

In producing this chapter I have been adopting a role not unlike that of devil's advocate. It need hardly be said that my intention has not been to diminish the undeniable power of the Jain vision so much as as to suggest how, in common with all developed religious traditions, it inevitably has built-in complexities that have to be taken into account lest a simplistic version of its message be constructed.[50] I have suggested how it is by no means straightforward, even for those who would locate Jainism exclusively in its most ancient manifestation, to use the early Jain scriptural tradition, as in the case of the story of "The Fine Horses," as an unambiguous source for environmentalism, and also that the nuances in respect to doctrinal issues shown by medieval writers must be taken into account by those wishing to understand Jainism fully.

While it might well be the case that Jainism has throughout its history developed certain insights that could be profitably deployed within the environmentalist debate, I am, broadly speaking, unhappy with the attempt, however well-intended it may be, to compel a traditional soteriological path originally mapped out for world-renouncing ascetics to fit the requirements of modern, ultimately secular, Western-derived agenda. Contemporary environmentalism seems to me to be a particular issue into which Indian religious traditions are coopted somewhat uneasily if their own often highly ambivalent presuppositions about nature and the world are not fully taken into account.[51]

Notes

1. For the *īryāpathikī* confessional formula, see R. Williams, *Jaina Yoga: A Survey of the Mediaeval Śrāvakācāras* (London: Oxford University Press, 1963), 203–4.

2. L. M. Singhvi, *The Jain Declaration on Nature*, p. 1. Reprinted in the appendix to this volume.

3. See Frank van den Bossche, "Jain Arguments against Vedānta Monism: A Translation of the *Parabrahmotthāpanasthala* of Bhuvanasundara Sūri," *Journal of Indian Philosophy* 25 (1997): 337–74. Cf. Lance E. Nelson, "The Dualism of Nondualism: Advaita Vedānta and the Irrelevance of Nature," in *Purifying the Earthly Body of God: Religion and Ecology in Hindu India*, ed. Lance E. Nelson (Albany: State University of New York Press, 1998), 61–88, for a compelling critique of the view that nondualism necessarily entails a reverence for nature.

I have recently drawn attention to some odd or ill-informed views of Jainism to be found in recent English-language fiction. See Paul Dundas, "The Laicisation of the Bondless Doctrine: A New Study of the Development of Early Jainism," *Journal of Indian Philosophy* 25 (1997): 511, n. 5. Another curious example worth noting in the present context, given the eminence of its author, is John Updike, *Toward the End of Time* (London: Hamish Hamilton, 1998), 34: "Today the sages say, via such Jainist cosmogonies as string theory and the inflationary hypothesis, that everything is nothing. The cosmos is a free lunch, a quantum fluctuation."

4. This judgment is found in chapter 19 of the *Prajñāpanā Sūtra*, an important survey of ontological and natural categories perhaps written some time around the turn of the common era. See *Paṇṇavaṇāsuttaṃ*, ed. Muni Puṇyavijaya, Pt. Dalsukh Malvaṇiā and Pt. Amṛtlāl Mohanlāl Bhojak, Jain Āgama Series, vol. 9, part 2 (Bombay: Shrī Mahāvīra Jain Vidyālaya, 1971), introduction, p. 361. For a general treatment of the subject, see J. C. Sikdar, "The Fabric of Life as Conceived in Jaina Biology," *Sambodhi* 3 (1974): 1–10.

5. Jain epistemology posits five types of knowledge or consciousness: direct sensory experience, reasoning, supersensory knowledge, clairvoyance, and omniscience. The final stage of omniscience, literally, "unique, isolated knowledge" (*kevalajñāna*), is envisaged in traditional Jainism as involving the simultaneous "seeing and knowing" of all the contents of the universe past, present, and future. This is achieved by eliminating the "harming" *karma*s which obstruct the soul's perception, knowledge, and energy. For the unenlightened who can only deploy the lower forms of knowledge, it is the scriptures, regarded as the words of the omniscient teachers, which mediate a correct, albeit indirect, understanding of the nature of the universe.

6. For some examples, see Padmanabh Jaini, "Indian Perspectives on the Spirituality of Animals," in *Buddhist Philosophy and Culture: Essays in Honour of N. A. Jayawickrema*, ed. D. J. Kalupahana and W. G. Weeraratne (Colombo: N. A. Jayawickrema Felicitation Volume Committee, 1987), 169–78.

7. The classical scriptural enunciation of this is found in the *Titthogālī*, one of the "mixed" (*prakīrṇaka*) category of scriptures of the Śvetāmbara Jain canon. For the text, see Muni Shri Punyavijaya and Pt. Amritlal Mohanlal Bhojak, *Paiṇṇayasuttāiṃ: Part 1*, Jain Āgama Series, no. 17, part 1 (Bombay: Shrī Mahāvīra Jaina Vidyālaya, 1984), 409–523. Compare also Dalsukh Malvania, "Study of Titthogaliya," in

Bhāratīya Purātatva: Purātatvācārya Muni Jinavijaya Abhinandan Granth (Jayapur: Śrī Muni Jinavijaya Sammān Samiti, 1971), 129–38.

8. This might be regarded as a kind of supplementary axiom to be added to the three formulated by Axel Michaels in a recent survey of traditional Indian views on nature. See Axel Michaels, "La Nature pour la Nature—Naturzerstörung und Naturschonung im traditionellen Indien," *Asiatische Studien/ Études Asiatiques* 50 (1996): 818. Michaels's axioms are as follows:

1. Not only does man destroy nature, but nature destroys itself cyclically.
2. Because man is part of nature he must also destroy himself.
3. Not only can nature be liberated, but so also can man, to the extent that he is involved in activity hostile to nature.

These axioms represent alternative positions to current ecological attempts to solve the problem of the degradation of the natural world and provide a valuable point of orientation for anyone intending to pronounce upon the relationship between nature and Indian religions without exclusive dependence upon Western categories or perspectives.

9. There was no complete unity among the Jain community about the propensities of the *kevalin*, the enlightened individual. For the Digambara sect and certain Śvetāmbaras, such as Dharmasāgara (sixteenth century), the *kevalin* could not even involuntarily act in such a way as to injure living creatures. For the great Śvetāmbara intellectual Yaśovijaya (for whom, see below), on the other hand, this would be to deny an inevitable aspect of the *kevalin*'s humanity.

10. Śivārya, *Bhagavatī Ārādhanā*, Bhāg 1, ed. Pt. Kailāścandra Siddhāntaśāstrī (Śolāpur: Jain Saṃskṛti Saṃrakṣak Saṅgh, 1978), vv. 87, 92, and 800. Śivārya may have belonged to the Yāpanīya sect.

11. Yuvācārya Mahāprajña, *Ahiṃsā Tattva Darśan* (Cūrū, 1988), 31.

12. This is the perspective of Rita M. Gross, "Buddhist Resources for Issues of Population, Consumption and the Environment," in *Buddhism and Ecology: The Interconnection of Dharma and Deeds*, ed Mary Evelyn Tucker and Duncan Ryūken Williams (Cambridge, Mass.: Harvard University Center for the Study of World Religions, 1997), 291: "As is the case with all major traditions, conclusions relevant to the current situation cannot be quoted from the classic texts; rather, the *values* inherent in the tradition need to be applied to the current, unprecedented crises of overpopulation and excessive consumption that threaten to overwhelm the biosphere upon which we are dependent."

13. For a recent attempt, not without its own interest, to present Jainism as a type of science, see K. V. Mardia, *The Scientific Foundations of Jainism* (Delhi: Motilal Banarsidass, 1996).

14. I have used the text of the *Sūtrakṛtāṅga* with Śīlāṅka's commentary in Muni Jambūvijaya's reedition, *Ācārāṅgasūtram and Sūtrakṛtāṅgasūtram* (Delhi: Motilal Banarsidass, 1978), 269–70.

15. After delivering a draft version of this paper at the Harvard conference, I was delighted to discover that Professor W. B. Bollée had recently completed a translation and commentary on *Sūtrakṛtāṅga* 2.6, which he kindly made available to me prior to its eventual publication as "Adda or the Oldest Extant Dispute between Jains and Heretics (Sūyagaḍa 2, 6): Part Two," *Journal of Indian Philosophy* 27 (1999): 411–

37. Professor Bollée believes that the fact that elephant meat is an inedible substance for both Hindus and Buddhists is a relevant factor here, and he suggests that the elephant ascetics may be an invention on the part of whoever wrote *Sūtrakṛtāṅga* 2.6 "as the climax of a series of abominable unbelievers."

I would add to this the possibility that the Jain author may be erroneously recording as a form of deviant religiosity the behavior of some sort of hunter-gatherer group living in the Ganges basin, not dissimilar to the Bushmen and Pygmies in Africa who only kill what they need and share it among members of the group. Cf. Hanns-Peter Schmidt, "Ahiṃsā and Rebirth," in *Inside the Texts, Beyond the Texts: New Approaches to the Study of the Vedas*, ed. Michael Witzel (Cambridge, Mass.: Department of Sanskrit and Indian Studies, Harvard University, 1997), 220.

However, as my colleague Dr. Alan S. Barnard (Department of Social Anthropology, University of Edinburgh) has pointed out to me (personal communication), hunter-gatherers, whether in Africa or elsewhere, do not plan ahead and store food, as seems to be implied by the behavior of the elephant ascetics. Instead, they live by "immediate return," and the sharing of the food, rather than any storage of it, is a vital part of this ideology.

16. As I stress throughout this chapter, a basic problem for those who would construct a Jain environmentalism is the ascetic preoccupations of the Jain texts upon which they draw and the consequent inappropriateness of artificially compelling them to fit the needs of an alternative context. The "behavior of the bees," referred to by Śīlāṅka, is simply an analogy for the calm, uninvolved obtaining of alms by monks and is an idiom found also in Hindu tradition, with no evidence for it having been borrowed from the Jain milieu. See Patrick Olivelle, *Rules and Regulations of Brahmanical Asceticism: Yatidharmasamuccaya of Yādava Prakāśa* (Albany: State University of New York Press, 1995), 449 (index: "Begging like a bee").

The *Jain Declaration on Nature* of 1992, however, elevates this usage into an ecological principle supposedly followed by the entire Jain community, according to which "[i]n their use of the earth's resources, Jains take their cue from the bee that sucks honey in the blossoms of a tree without hurting the blossom and strengthens itself." Here is a good example of an ancient scriptural idiom of relevance purely to the monastic community being decontextualized and fitted into a modern ecological ethic.

It might also be argued that we are not too far here from the contemporary phenomenon that scholars of Buddhism have called the "monasticiation of the laity," whereby in certain modern contexts the laity's perception of religious activity is strongly informed by texts, with the result that they sometimes structure their lives in ways replicating those of monks. This approach might be inevitable in the context of the Jain diaspora in the West where there are few, if any, representatives of ascetic authority, but it flies in the face of a substantial element of Jain tradition.

17. See Paul Dundas, *The Jains* (London and New York: Routledge, 1992), 199; W. J. Johnson, "The Religious Function of Jain Philosophy: Anekāntavāda Reconsidered," *Religion* 25 (1995): 41–50; and Jayandra Soni, "Philosophical Significance of the Jaina Theory of Manifoldness," *Studien zur interkulturellen Philosophie* 7 (1997): 277–87.

18. See the discussion in Phyllis Granoff, "The Violence of Non-Violence: A

Study of Some Jain Responses to Non-Jain Religious Practices," *Journal of the International Association of Buddhist Studies* 15 (1992): 1–45.

19. For example, the Jains who work at the bottom level of the Jaipur diamond trade or the Digambara peasants on the Maharashtra-Karnataka border.

20. See K. R. Norman, "The Role of the Layman according to the Jain Canon," in *The Assembly of Listeners: Jains in Society*, ed. Michael Carrithers and Caroline Humphrey (Cambridge: Cambridge University Press 1991): 31–39.

21. See Muni Jambūvijaya's edition, Nāyādhammakahāo (Jñātādharmakathāṅgasūtram), Jaina Āgama Series, vol. 5, part 1 (Bombay: Shri Mahavira Jaina Vidyālaya, 1989), 312–33 (*sattarasamaṃ ajjhayaṇaṃ "Āiṇṇe"*); and for a brief German summary, see Walther Schubring, *Nāyādhammakahāo: Das sechste Anga des Jaina-Siddhānta* (Mainz: Akademie der Wissenschaften und der Literatur, 1978), 59–60.

22. See Paul Dundas, "The Digambara Jain Warrior," in *The Assembly of Listeners*, ed. Carrithers and Humphrey, 181–82; and John E. Cort, "Who Is a King? Jain Narratives of Kingship in Medieval Western India," in *Open Boundaries: Jain Communities and Cultures in Indian History*, ed. John E. Cort (Albany: State University of New York Press, 1998), 85.

23. *Nāyādhammakahāo*, p. 328, lines 6–9, and p. 329, lines 12–15.

24. See Williams, *Jaina Yoga*.

25. Muṃbaī: Jinaśāsana Ārādhanā Ṭrasṭ, v. s. 2044.

26. See pp. 1a and 26b:

vavahārasuddhi dhammassa mūlaṃ savvannu bhāsae,
vavahāreṇaṃ tu suddheṇaṃ atthasuddhī jao bhave.
suddheṇaṃ ceva attheṇaṃ āhāro hoi suddhao,
āhāreṇaṃ tu suddheṇaṃ dehasuddhī jao bhave.
suddheṇaṃ ceva deheṇaṃ dhammajuggo a jāyaī,
jaṃ jaṃ kuṇai kiccaṃ tu taṃ taṃ se saphalaṃ bhave.

27. Page 1a:

na hi tad vidyate kiṃ cid yad arthena na siddhyati
yatnena matimāṃs tasmād artham ekaṃ prasādhayet.

28. Page 28a:

lokaḥ khalv ādhāraḥ sarveṣāṃ dharmacāriṇāṃ yasmāt,
tasmāl lokaviruddhaṃ dharmaviruddhaṃ ca saṃtyājyam.

The specific ascription of the verse is to "*vācakamukhya*," i.e., Umāsvāti (beginning of the common era). The source is *Praśamaratiprakaraṇa*, v. 131. I refer to *Vācaka Śrīmad Umāsvāti's Praśamarati-prakaraṇa*, trans. and appendices in English by Mahesh Bhogilal, ed. V. M. Kulkarni (Ahmedabad: Mrs. Nita M. Bhogilal et al., 1989), 50.

29. See Williams, *Jaina Yoga*, 117.

30. D. N. Bhargava, "Pathological Impact of [sic] Environment of Professions Prohibited by Jain Acaryas," in P. S. Jain and and R. M. Lodha, *Medieval Jainism: Culture and Environment* (New Delhi: Ashish Publishing House 1990), 103–8.

31. *Vyavahāraśuddhiprakāśa*, p. 1b, and cf. Williams, *Jaina Yoga*, 121–22.

32. Williams, *Jaina Yoga*, 118.

33. *Vyavahāraśūddhiprakāśa*, p. 2a.

34. Mahāprajña, *Ahiṃsā Tattva Darśan*, p. 36.

35. This is not the occasion to discuss the dating of Haribhadra. However, Williams, *Jaina Yoga*, 6–7, is clear that Haribhadra's *Pañcāśaka*, to which I refer below, may be dated to the sixth century C.E. For the *Pañcāśaka* with Abhayadeva Sūri's commentary, I have used the 1912 edition published in Bhāvnagar by the Jaindharmaprasārakasabhā.

36. *Pañcāśaka*, 7.42, with Abhayadeva Sūri's commentary. Cf. Haribhadra, *Ṣoḍaśaka*, ed. Buddhisāgara, Devcandra Lālbhāī Jainpustakoddhār, vol. 6 (Muṃbaī, 1911), 6.16.

37. See *Ṣoḍaśaka*, 6.7–8, for wood used for this purpose, and cf. Abhayadeva Sūri on *Pañcāśaka* 7.17.

Deforestation has probably been the most significant ecological problem in South Asia throughout its history. The one significant Jain text where we do find extensive discussion of the desirability of not harming trees, namely, the *Niśītha Sūtra* (ed. Yuvācārya Madhukar Muni [Byāvar: Āgamprakāśan Samiti, 1991], 256–62), is purely ascetic in orientation. The minutiae of descriptive detail there revealed display not so much concern for plant life for its own sake, but as a desire to emphasize to the ascetic the extent to which he has to be on guard if he wishes to advance on the path. Furthermore, if it still be suggested that such textual statements might provide the underpinning for a Jain environmental ethic, let it be recalled that what is numerically the major Śvetāmbara sectarian tradition, the Tapā Gaccha, holds to this day that only male ascetics and no one else can read and act upon disciplinary texts such as the *Niśītha Sūtra*.

Medieval Indian scholars such as Sūrapāla developed a minute understanding of trees and the conditions under which they flourished, and dendrology (*vṛkṣāyurveda*) represented a serious and elaborate branch of traditional knowledge. See Rahul Peter Das, *Das Wissen von der Lebenspanne der Bäume: Sūrapāla's Vṛkṣāyurveda kritisch ediert, übersetzt and kommentiert* (Stuttgart: Franz Steiner, 1988). The existence of such a *śāstra* should not, of course, lead to speculation about some premodern, uniquely Indian connection with the natural world. Trees were not studied simply for botanical or ecological reasons but as part of the more broadly developed practical science of agriculture. Furthermore, knowledge of trees and their varieties and habitat was prized, by the Jains no less than other Indian sectarian communities, because they were perceived as providing markers for treasure hunters seeking wealth buried in the ground. For this, see Nalini Balbir, "A la recherche des trésors souterrains," *Journal of the European Āyurvedic Society* 3 (1993): 15–55.

38. *Pañcāśaka*, 7.37-41.

39. To be precise, it is *bhāṣya*, verse 194, in the cluster of verses following *Āvaśyaka Niryukti* 1056. See *Āvaśyaka Niryukti*, bhāg 2 (Muṃbaī: Śrī Bherulāl Kanaiyālāl Koṭhārī Dhārmik Ṭrasṭ v. s. 2038), 2, which gives the following text: *akasiṇapavattayāṇaṃ virayāvirayāṇaṃ esa khalu jutto / saṃsārapayaṇukaraṇo davvatthae kūvadiṭṭhaṃto*. The version given by the *Pañcāśaka* has slightly different readings, most notably *saṃsārapayaṇukaraṇe*. See Dundas, *The Jains*, 254, n. 4, for some other occurrences of this analogy.

40. *Āvaśyaka Niryukti*, p. 2.

41. *Pañcāśaka*, p. 120b: *tatprayogaś caivam—sadoṣam api svarūpeṇa yad guru-*

guṇāntarakāraṇaṃ tad āśrayaṇīyaṃ, yathā kūpakhananam, tathā ca dravyastava iti dṛśyam.

42. See Haribhadra, *Pañcavastuka* (Bhuleśvar: Jinaśāsana Ārādhanā Ṭrasṭ, v. s. 2045), v. 101, autocommentary.

43. I have used the edition of the *Kūpadṛṣṭāntaviśadīkaraṇa* published in a collection of five works (*Paṃcagranthi*) by Yaśovijaya, edited by Acharya Shri Yaśodevasurishvaraji Maharaj (Bombay: Shri Yasobharati Jain Prakashan Samiti, 1980), 66–94. There is another edition of the work in a collection of six works by Yaśovijaya entitled *Lokanālidvātriṃśikādiprakaraṇaṣaṭkam*, edited by Vijayajinendra Sūri, Harṣapuṣpāmṛta Jain Granthamālā, vol. 266 (Śāntipurī, 1993), 41–51.

44. *Kūpadṛṣṭāntaviśadīkaraṇa*, p. 69, commenting on verse 2, a modified version of *Āvaśyaka Niryukti bhāṣya*, verse 104.

45. *Kūpadṛṣṭāntaviśadīkaraṇa*, p. 70: . . . *dharmārthapravṛttāv apy ārambhajanitasyālpasya pāpasyeṣṭatvāt.*

46. *Kūpadṛṣṭāntaviśadīkaraṇa*, p. 70.

47. *Kūpadṛṣṭāntaviśadīkaraṇa*, vv. 3 and 4.

48. *Kūpadṛṣṭāntaviśadīkaraṇa*, p. 72:

> *dehādinimittaṃ pi hu je kāyavahaṃmi payaṭṭanti*
> *jiṇapūyākāyavahaṃmi tesiṃ apavattaṇaṃ moho.*
> This verse is *Pañcāśaka*, 4.45.

49. *Kūpadṛṣṭāntaviśadīkaraṇa*, v. 12: *ārambho vi hu eso haṃdi aṇārambho 'tti ṇāyavvo.*

50. For another example of this in the context of food, in the Jain imagination the most dangerous manifestation of the natural world for human beings, see Paul Dundas, *The Meat at the Wedding Feasts: Kṛṣṇa, Vegetarianism and a Jain Dispute*, The 1997 Roop Lal Jain Lecture (Toronto: University of Toronto, Centre for South Asian Studies, 1997).

51. Rather than continually invoking the more technical ontological areas of the religion, those wishing to elaborate an accessible environmental ethic grounded upon Jainism might do well to consider tapping the rich resources of medieval Jain narrative literature as a possible source of exemplification. Let me here simply refer to the eleventh-century Devacandra Sūri's story of Ārāmaśobhā, "The Girl Adorned with a Garden."

This young Brahmin girl was, on the death of her mother, obliged to carry out all the household chores for her father and stepmother. One day, tormented by the heat of the sun as she was looking after the family cows, she rescued a snake that was being pursued by snake charmers. The snake had in actuality been possessed by a semidivine serpent prince who, through his magic power, granted the exhausted girl the boon of a delightfully cool grove that would follow her wherever she went. In this story protection of one of nature's creatures leads to the reward of immersion in a benign omnipresent natural world. For a translation of the whole story, see Phyllis Granoff, trans., *The Forest of Thieves and the Magic Garden: An Anthology of Medieval Jain Stories* (Delhi: Penguin Books, 1998), 264–91.

The Living Earth of Jainism and the New Story: Rediscovering and Reclaiming a Functional Cosmology

CHRISTOPHER KEY CHAPPLE

In various chapters of this book, several authors have asserted that the Jain practice of nonviolence provides a firm foundation for the development of an environmental ethic. Citing the examples of appropriate livelihood, strict vegetarianism, and holistic logic as found within the Jain tradition, Padmanabh Jaini, Sadhvi Shilapi, Kim Skoog, John Koller, and others have seen parallels between the Jain concern for not harming life in all its various forms and the ethos of environmental protection.

In this chapter, I want to take a somewhat more metaphysical (or perhaps physical or biological) approach to interpreting Jainism in light of ecological exigencies. Jain history and sociology have demonstrated for centuries an unusually tenacious commitment to a very rigorous ethical system. It seems important to examine closely the underlying worldview from which the Jain practices of nonviolence, truthfulness, and so forth, arise. Specifically, this chapter will focus on two primary aspects of Jain teachings in light of three contemporary Western ecological thinkers. The first aspect of Jainism to be discussed is its unique cosmology, which will be compared to the cosmological insights of contemporary science as presented by Brian Swimme. The second aspect of Jainism to be explored is the Jain assertion that the seeming inert, nonsensate world abounds with sensuousness. The Jains posit that all the myriad living beings, from a clod of dirt or a drop of water to animals and humans themselves, possess one commonality: the capacity for tactile experience. This "living world" perspective will be discussed in light of Thomas Berry's call for understanding the earth as a "communion of subjects not a collection of objects." Furthermore, Jainism, in addition to positing a world populated with infinite life-forms in a wide range of manifestations, also develops an elaborate biological systemization of life that pays close attention to the role of the senses. As life-forms complexify, they add additional senses. By examining the implications of this polysensate view of the world and by exploring the underlying motivations for perceiving the world as suffused with life, and hence worthy of our respect and care, a glimpse into the emotionality underlying the Jain commitment to nonviolence might be gained. This approach will be compared with the writings of David Abram, a phenomenologist and philosopher who extols the role of the senses in developing an appreciation for the natural world.

The Jain perspective that the manifold parts of the world, including the elements themselves, contain "touch, breath, life, and bodily strength"[1] will be compared with the scientific view of the universe's dynamism as summarized by contemporary cosmologist Brian Swimme. The implications of the Jain panpsychic vision and sensibility will be juxtaposed with Thomas Berry's plea for increased sensitivity to the earth community, as embodying differentiation, subjectivity, and communion. In a third and final section, the very sensuousness and vitality of Jain philosophy will be discussed in the context of philosopher David Abram's appeal for a deeper appreciation of human reciprocity through the senses with the things of the world.

The method that I employ here is one of creative juxtaposition. Jain cosmology does not fundamentally share the same story as that put forth by contemporary science. Though Jainism emphasizes the status and condition and purification of the human soul as its primary concern, this does not match up, point by point, with Thomas Berry's call for increased subjectivity. Additionally, Abram's philosophy of the integration of body, mind, and landscape does not parallel Jainism's view of repeated rebirth, according to the laws of *karma*, until one achieves release. However, I introduce the deliberations of these three contemporary thinkers because they raise issues of the relationship between the body, consciousness, and the world in an attempt to cultivate greater sensitivity to the larger order and intricacy of the universe. They also raise questions of relationship to and responsibility for the natural order that call for the development of an ethical stance of awareness, care, and protection—an ethical stance not too far distant from the nonviolence advocated by the Jains. As the contemporary world seeks to understand traditional indigenous values that cultivate a respect for nature, and as traditions such as Jainism seek to maintain relevance in the modern world, such points of dialogue as posed in this volume and in this essay might be helpful, not only for intercultural understanding, but also to advance the shared goal of preserving and respecting all forms of life.

Jain Cosmology: A Universe Permeated with Life

Stories of cosmology ground the human person within the world. They explain the place of the individual within the larger context of social and physical realities. In ancient India as articulated in the *Ṛgveda*, the person, or *puruṣa*, was regarded as a reflection of the world itself in its great immensity: eyes were said to correspond to the sun; the mind was correlated with the moon; breath with the wind; feet with the earth. This particular cosmology asserts a linkage between the microphase and the macrophase; by seeing the universe as reflective of and relating to body functions, one sees oneself not as an isolated unit but as part of a greater whole. The Jain tradition developed a parallel story of the structure of the cosmos, complete with the image of a great female whose body symbolizes the entire system. However, whereas the texts of the early Vedic tradition remain somewhat vague about the place of individual life force in this process, Jainism develops an intricate accounting for the journey for each life force (soul, or *jīva*), which is said to be eternal, not created by any deity, and ultimately responsible for its own destiny. In this section of the chapter, Umāsvāti's explication of traditional Jain cosmology, which dates from the early centuries of the common era, will be compared and contrasted with Brian Swimme's explication of contemporary cosmological science.

Jainism provides one of India's most thorough attempts to encapsulate a comprehensive worldview or cosmology that integrates the place of the human person within the continuum of the universe.

Umāsvāti's system is accepted by both major branches of Jainism, the Digambara and the Śvetāmbara. It attempts to explain the place of the human being in a great continuous reality. It further, as mentioned above, emphasizes hierarchy and vitality within its vision of the cosmos.

Jain cosmology describes a storied universe in the shape of a female figure. The earthly realm, or middle world (*manuṣya loka*), consists of three continents and two oceans. Animals (as listed below), including humans, can be found there. Below the earth can be found seven hells. Above earth, eight heavenly realms are arrayed. The ultimate pinnacle of the Jain system, symbolized at the top of the head of the cosmic person, consists of the state of liberation, the *siddha loka*. Human beings who have successfully led a religious life achieve this through the release of all karmic bondage. One cannot attain this state from the heavenly or hellish realms; only through a human birth lived well according to spiritual precepts can this final abode be gained. According to Umāsvāti's *Tattvārtha Sūtra*, 8,400,000 different species of life-forms exist.[2] These beings are part of a beginningless round of birth, life, death, and rebirth. Each living being houses a life force, or *jīva*, that occupies and enlivens the host environment. When the body dies, the *jīva* seeks out a new site depending upon the proclivities of *karma* generated and accrued during the previous lifetime. Depending upon one's actions, one can either ascend to a heavenly realm, take rebirth as a human or animal or elemental or microbial form, or descend into one of the hells, as a suffering human being or a particular animal, depending upon the offense committed.

The taxonomy of Jainism, which will be discussed in greater detail in the next section of this chapter, places life-forms in a gradated order starting with those beings that possess only touch, the foundational sense capacity that defines the presence of life. These include earth, water, fire, and air bodies; microorganisms (*nigoda*); and plants. The next highest order introduces the sense of taste; worms, leeches, oysters, and snails occupy this phylum. Third-order life-forms add the sense of smell, including most insects and spiders. Fourth-level beings, in addition to being able to touch, taste, and smell, also can see; these include butterflies, flies, and bees. The fifth level introduces hearing and is further divided into categories of those nonsentient and sentient. Birds, reptiles, mammals, and humans dwell in this life realm.[3]

Jainism posits a cosmological view that at first glance seems similar to that put forth in Ptolemy's theory of the spheres and Dante's *Divine Comedy*. At the base of this cosmos can be found various regions of hell. In the central realm is the surface of the planet, on which reside the five elements, living beings, and humans. Above this realm extends a sequence of heavenly worlds. At the pinnacle of this cosmos exists a domain of liberated beings who have risen above the vicissitudes of repeated birth in the lower, middle, and higher realms. In spatial orientation and its theory of moral consequences, it seems to evoke Dante's system of hell, purgatory, and heaven. Depending on one's actions, one earns a berth in one of the three domains.

However, if we look more closely at this system, its theories of space, time, and matter carry more subtlety and sophistication than may first seem apparent. First, Jainism identifies two primary categories of reality: living and nonliving. Living reality, or *jīva*, is broadly defined as dynamism and suffuses what in precontemporary physics would be considered inert. Each *jīva* is said to contain consciousness, energy, and bliss. Earth, water, fire, air bodies, which comprise material objects such as wood or umbrellas or drops of water or flickers of flame or gusts of wind all contain *jīva*, or individual bodies of life force. The category of nonliving "things" includes properties such as the flow of time and space and the binding of matter known as *karma*, or *dravya*, onto the *jīva*. The nature of this *karma* determines the course of one's embodiment and experience. Negative *karma* causes a downward movement both in this birth and in future birth. Positive *karma* releases the negative, binding qualities of *karma* and allows for an ascent to higher realms, either as a more morally pure human being or as a god or goddess. Ultimately, the Jain path of purification through its many strict ethical precepts may culminate in joining the realm of the perfected ones, the *siddhas*. These liberated souls have released themselves from all *karma* and dwell in a state of eternal consciousness, energy, omniscience, and bliss.

In this cosmological system, one's station in life can be understood in terms of one's degree of effort in following ethically correct patterns of life as taught by the Jain Tīrthaṅkaras, or spiritual leaders. The world of nature cannot be separated from the moral order; even a clod of earth exists as earth because it has earned its particular niche in the wider system of life processes. A human's experience includes prior births of various animals, microorganisms, elemental entities, and perhaps gods and goddesses. To see and recognize and understand the world is to acknowledge one's past and potential future. Though the Jain insistence on the uniqueness of each individual soul does not lend itself to an ultimate vision of interconnected monism, it nonetheless lays the foundation for seeing all beings other than oneself with an empathetic eye. In past or future births, one could have been or could become a life-form similar to any of those that surround one in the vast unlimited cosmos.

The Story of Contemporary Cosmology

The contemporary story of the universe as told by physicists and cosmologists is complex and varied, requiring an understanding of higher mathematics and a reliance on sophisticated instruments, such as electron microscopes, and telescopes that penetrate deep into distant galaxies. Though many interpreters of science, such as Stephen Hawkins and Carl Sagan, have summarized the various theories about the origins and structure of the universe, few have attempted to create a world of meaning from this raw data. Brian Swimme, however, has attempted to make sense of the insights of modern physics and examine the implications of this newly discovered world order for human behavior. In this section, one aspect of his interpretation will be summarized and then discussed in light of Jainism and the larger context of environmental ethics.

In their observations of the behavior of matter and energy, planets and galaxies, Einstein and Hubble calculated that the world flared forth some fifteen billion years ago. From that time and point of origin, all things blasted forth away from one another. The stuff of stars and elements continue to move apart from one another and, over the course of fifteen billion years, as yet uncounted galaxies spin forth and continue to move outward. Because of this initial momentum, everything retains a part of this original being. And because everything continues to move from that point of origin, everything that contains a bit of that point of origin is at the center of everything else that is moving forth. And because everything is moving forth and everything originated from that original flaring moment, everything is the center of the universe and yet is moving from everything else.

Furthermore, the space that separates all these discrete masses of atomic materiality continues to generate evanescent particulate matter that constantly emerges and then dissolves. Even empty space is not empty but carries what Swimme describes as the "all-nourishing abyss." As he describes it:

> The usual process is for particles to erupt in pairs that will quickly annihilate each other. Electrons and positrons, protons and anti-protons, all of these are flaring forth, and as quickly vanishing again. Such creative and destructive activity takes place everywhere and at all times throughout the universe. The ground of the universe then is an empty fullness, a fecund nothingness. Even though this discovery may be difficult if not impossible to visualize, we can nevertheless speak a deeper truth regarding the ground state of the universe. First of all it is not inert. The base of the universe is not a dead, bottom-of-the-barrel thing. The base of the universe seethes with creativity, so much so that physicists refer to the universe's ground state as 'space-time foam.'"[4]

This account of materiality abounds in mystery, unpredictability, and dynamism. The ground for the manifested world lies hidden in forces like the yin and yang of Chinese philosophy that constantly vacillate between presence and absence. Furthermore, like the Jain system of transmutation of life-forms, this primal energy constantly seeks new expression.

Both the story of contemporary cosmology and that of Jainism allow for awe and respect for materiality. According to Swimme, our deadened view of the material has led to the blight of consumerism, where ultimate meaning in life is mistakenly sought in the accumulation of things. This has resulted in lives of loneliness, depression, and alienation. He writes:

> Consumerism is based on the assumption that the universe is a collection of dead objects. It is for this reason that depression is a regular feature in every consumer society. When humans find themselves surrounded by nothing but objects, the response is always loneliness. . . ."[5]

For Swimme, the remedy for this angst can be found in a rediscovery of awe through appreciation of the intricacy and beauty of the material world, from the complexity of the meadow to the splendid grandeur of the Milky Way. Swimme writes that

> Each person *lives* in the center of the cosmos. Science is one of the careful and detailed methods by which the human mind came to grasp

> the fact of the universe's beginning, but the actual origin and birthplace is not a scientific idea; the actual origin of the universe is where you live your life. . . . 'The center of the cosmos' refers to that place where the great birth of the universe happened at the beginning of time, but it also refers to the upwelling of the universe as river, as star, as raven, as you, the universe surging into existence anew.[6]

In this vision of the human place within the cosmos, each individual, each context holds ultimate meaning in its immediacy and its ongoing participation in the process of cocreation. As centers of creativity, all beings, all particles, play an important, integral role in the greater scheme of things. While retaining a unique and unencroachable perspective, each point of life holds a commonality with all others due to their shared moment of origin fifteen billion years ago.

In some ways, this vitalistic account of creation and reality bears similarities with the Jain tradition; there are also notable differences. The fundamental disagreement would lie in the premise that the world began in the single moment of the Big Bang or Flaring Forth.[7] Jainism, like Buddhism, asserts the eternality of the universe and rejects the notion of an initial creation moment. However, just as Swimme contends that the consumerist obsession with "dead" objects leads to depression, in Jainism, the abuse and manipulation of materiality leads to a thickening of one's karmic bondage, guaranteeing a lower existence in this and future lives. Swimme suggests that the things of the world be regarded as a celebration of the originary moment of creation, that people turn their attention to the beauty and mystery of creation as an antidote to the trivialization of life brought about by advertisements and accumulations. Jainism similarly asserts that things share a commonality in their aliveness, which must be acknowledged and protected. Through respect for life in all its forms, including microorganisms and the elements, one can ascend to a higher state of spiritual sensitivity.

Traditional Jain cosmology and contemporary scientific accounts of the workings of the universe hold implications for the development of ecological theory. Both systems place value on the natural order. Both systems hold the potential to evoke the affective dimension of human responsiveness. Both systems develop an ethical view that calls for greater awareness of one's immediate ecological context. Swimme's system offers a prophetic critique of unbridled consumerism and its consequent trivialization and deadening of the material world. Jainism develops a specific code of behavior that seeks to respect the life force in its various forms, including its material manifestations.

Swimme's summary explanations of contemporary cosmology present the central notions of Hubble's cosmological discoveries in a succinct and poignant manner, not unlike the Sūtra style employed by Umāsvāti to provide a Jain account for the structure of reality. These two systems presented by Umāsvāti and Swimme carry an inherent ethical and perhaps teleological message. Jainism explains the universe so that it can make sense of its theology of spiritual liberation. Swimme explains the universe in an attempt to wrest humans from their blind allegiance and devotion to a numbing materialistic view that regards the things of the universe as dead and inert. Both provide an occasion to view the world as a living, dynamic process that, in the contemporary context of environmental degradation, requires protection and care. In the next section of this chapter we will investigate how the particularities of Jain biology might be used to enhance one's sense of the universe as a living process of multiple subjectivities rather than as a chaotic assemblage of inert materiality.

The Hierarchy of Life in the Jain Tradition

The *Ācārāṅga Sūtra*, the earliest known Jain text, describes a world suffused with life. In relating the life story of Mahāvīra, the twenty-fourth great teacher, or Tīrthaṅkara, who lived in the fourth or fifth century B.C.E., the text states that

> Thoroughly knowing the earth-bodies and water-bodies and fire-bodies and wind-bodies, the lichens, seeds, and sprouts, he comprehended that they are, if narrowly inspected, imbued with life. . . .[8]

From this perception of the livingness of all things as articulated by Mahāvīra, Jainism developed an extensive theory of *karma* to account for the existence of various life-forms. According to Jain *karma* theory, each life-form will eventually take on a new existence as part of the ongoing process of *saṃsāra*, to be halted only when one, as a human being, attains spiritual liberation *(kevala)*.

Mahāvīra laid out a series of rules to assist one along the path to liberation. These rules were designed to minimize and eliminate *karma* through a careful observance of nonviolent behavior. Mahāvīra instructs his monks and nuns to avoid harming life in its myriad forms through various methods. These include explicit instructions for when and what and how to eat; when and how to travel; where and when to defecate; from whom to accept food; and lists of various other activities, including attendance at wedding ceremonies, to be avoided.[9] All these rules, as well as the various preferred professions for laypersons, which have been mentioned in other chapters, are to be observed in order to prevent harm to living beings. In fact, Mahāvīra even exhorts his monks and nuns not to gesture or point because "The deer, cattle, birds, snakes, animals living in water, on land, in the air might be disturbed or frightened, and strive to get to a fold or . . . refuge, (thinking): 'the Sramana [monk] will harm me!'"[10] This profound respect for the natural world distinguishes Jainism among the world's religious traditions as potentially the most eco-friendly.

In later Jain literature, various authors describe the living world with a great deal of care and precision. For instance, Śānti Sūri, a Śvetāmbara Jain writer of the eleventh century, provides elegant descriptions of living beings, beginning with the earth beings and concluding with various classes of deities and liberated souls. In the *Jīva Vicara Prakaranam*, a text of fifty verses, he lists types of life, frequency of appearance, and cites an approximate lifespan for each. For instance, he states that hardened rock can survive as a distinct lifeform for twenty-two thousand years; "water-bodied souls" for seven thousand years; wind bodies for three thousand years; trees for ten thousand years; and fire for three days and three nights.[11] Each of these forms demonstrates four characteristics: life, breath, bodily strength, and the sense of touch.[12]

The attention to detail given to the elemental realm of one-sensed beings distinguishes the medieval Jains as closely observant scientists. Their descriptions include fundamental information regarding geology, meteorology, botany, and zoology. Śānti Sūri describes the one-sensed living realm with great precision, extending from the earth through water and fire and air to the plant kingdom. For the *prthivīkāyika jīva*s, or earth-bodied souls, he offers the following two verses:

> Crystalline quartz, jewels, gems, coral, vermilion, orpiment, realgar, mercury, gold, chalk, red soil, five-colored mica, hard earth, soda ash, miscellaneous stones, antimony, lava, salt, and sea-salt are the various forms taken by the earth-body souls (Prithivikayika Jivas).[13]

The numerous types of stone and soil listed indicate that the Jains were keen observers of geological formations, careful to distinguish the characteristics of color, density, and hardness.

Śānti Sūri's descriptions of the various forms of water are similarly perspicacious, listing:

> Underground water, rainwater, dew, ice, hail, water drops on green vegetables, and mist as the "numerous varieties of Water-bodied Souls."[14]

Śānti Sūri provides an exhaustive list of various forms taken by fire-bodied souls:

> Burning coals, flames, enflamed cow dung, fire reflected in the sky, sparks falling from a fire or from the sky, shooting stars, and lightning constitute Agnikaya Jivas.[15]

The various wind bodies are listed as follows:

> Winds blowing up, winds blowing down, whirlwinds, wind coming from the mouth, melodious winds, dense winds, rarefied winds are the different varieties of Vayu Kayika Jivas.[16]

Descriptions of various plant genres then follow, with precise detail given for plants with fragance, hard fruits, soft fruits, bulbous roots, thorns, smooth leaves, creepers, and so forth. Lists are offered to restrict or endorse the use of specific plants, with special attention paid to determining avoidance of undo harm to plants that harbor the potential for even greater production of life-forms.

Two-sensed beings, possessing touch and taste, are said to live twelve years and include conches, cowries, gandolo worms, leeches, earthworms, timber worms, intestinal worms, red water insects, white wood ants, among others.[17] Three-sensed beings live for forty-nine days and include centipedes, bedbugs, lice, black ants, white ants, crab lice, and various other kinds of insects.[18] These beings add the sense of smell. Four-sensed beings, which add the sense of sight, live for six months[19] and include scorpions, cattle bugs, drones, bees, locusts, flies, gnats, mosquitoes, moths, spiders, and grasshoppers.[20] At the top of this continuum reside the five-sensed beings, which add the sense of hearing and can be grouped into those who are deemed "mindless" and those who are considered to be sentient. This last group includes the denizens of hell, gods, and humans. Various life spans are cited for five-sensed beings, which Śānti Sūri describes in great detail: land-going, aquatic, sky-moving, and so forth. The detailed lists by Śānti Sūri and his later commentators present a comprehensive overview of life-forms as seen through the prism of Jainism.

The Jain worldview cannot be separated from the notion that the world contains feelings and that the earth feels and responds in kind to human presence. Not only do animals possess cognitive faculties including memories and emotions, the very world that surrounds us can feel our presence. From the water we drink, to the air we inhale, to the chair that supports us, to the light that illumines our studies, all these entities feel us through the sense of touch, though we might often take for granted their caress and support and sustenance. According to the Jain tradition, humans, as living, sensate, thinking beings, have been given the special task and opportunity to cultivate increasingly rarefied states of awareness and ethical behavior to acknowledge that we live in a universe suffused with living, breathing, conscious beings that warrant our recognition and respect.

Various authors within the Western biological, philosophical, and psychological disciplines have similarly argued for the possibility that animals possess cognition and that the world itself cannot be separated from our cognition of it. Few have committed themselves to the very radical Jain notion that the elements possess consciousness, though some environmental thinkers, such as Christopher Stone, have argued for the legal standing of trees. But, as discussed in the sections that follow, Thomas Berry and David Abram have argued that a heightened responsiveness to the earth is essential for the full development of human consciousness.

The New Story of Thomas Berry: A Call for Sensitivity to Life

Thomas Berry has advocated the telling of a "new story" that allows us to reinhabit the earth with a greater awareness of the fragile balance of life systems. He writes:

> The human species has emerged within this complex of life communities; it has survived and developed through participation in the functioning of these communities at their most basic level. Out of this interaction have come our distinctive human cultures. But while at an early period we were aware of our dependence on the integral functioning of

these surrounding communities, this awareness faded as we learned, through our scientific and technological skills, to manipulate the community functioning to our own advantage. This manipulation has brought about a disruption of the entire complex of life systems. The florescence that distinguished these communities in the past is now severely diminished. A degradation of the natural world has taken place.[21]

Berry suggests that, with the waning of traditional creation stories and functional cosmologies, we must develop a new story that can effectively replace them and introduce a new integrated worldview. This worldview must account for the workings of the universe, inspire awe at its grandeur, and prompt the earth's citizens to an appropriate response to enhance the sustainability of the earth. Drawing from the pioneering insights of the Jesuit geologist and theologian Pierre Teilhard de Chardin, Berry suggests an embrace of the cosmological story emerging from the new science. In his focus on the notion of a fixed point of creation and his orientation toward an almost eschatological prophetic voice, Berry's work seems well-grounded in the Jewish/ Christian/ Islamic tradition. Yet, in other ways, it is similar to and clearly informed by various aspects of Asian, African, and tribal traditions.

For the past twenty years, Thomas Berry has written and lectured on the topic of the emerging ecozoic age. Taking note of the tremendous harm caused to the environment during the twentieth century, he observes that we have lost touch with the natural world, that we have become callous to the magnificent universe that supports and nurtures us. During a plenary address to the American Academy of Religion in 1993, Berry stated:

We hardly live in a universe at all. We live in a city or nation, in an economic system, or in a cultural tradition. We are seldom aware of any sympathetic relation with the natural world about us. We live in a world of objects, not in a world of subjects. We isolated ourselves from contact with the natural world except insofar as we enjoy it or have command over it. The natural world is not associated with the very meaning of life itself. It is little wonder that we have devastated the planet so extensively.[22]

The causes of the rift between humans and nature are numerous, layered, and storied. As noted by Lynn White, Jr., the religious traditions of the West find their roots in an entrenched anthropocentrism that places emphasis on dominion over nature. As Berry has written, the concern with redemption in Western religious traditions leaves little room for an appreciation of the natural world, which is seen as subsidiary to the interests of human comfort. The exploitative mentality of New World settlement, the rise of industrialization in the eighteenth century, the explosion of consumerism and technology in the twentieth century propelled the human into a new relationship with nature. Berry writes:

> Here it is necessary to note that planet Earth will never again in the future function in the manner that it has functioned in the past. Until the present the magnificence splashed throughout the vast realms of space, the luxuriance of the tropical rainforests, the movement of the great whales through the sea, the autumn color of the eastern woodlands; all this and so much else came into being entirely apart from any human design or deed. We did not even exist when all this came to be. But now, in the foreseeable future, almost nothing will happen that we will not be involved in. We cannot make a blade of grass, but there is liable not to be a blade of grass unless we accept it, protect it, and foster it.[23]

We have entered into a new phase of earth-human relations, wherein the human effectively has conquered nature. The now-submissive earth relies upon the human for its continuance. The earth has been bruised by the abundance of radioactive waste and the ever-present threat of nuclear conflagration. The sky has been fouled with emissions from automobiles, scooters, and factories. Human and industrial wastes have polluted our rivers and lakes. Life itself has become imperiled.

The Realm of the Senses: The Experience of Life

As this separation takes place, humans lose their intimacy with the natural world and themselves. With this loss of intimacy comes a deadening indifference to the natural world, which results in further exploitation and destruction. To reverse this process, one needs to recapture a sense of beauty and appreciation for the natural world, a sense of the wholly real materiality of things, not for the sake of consumption and manipulation but for the very being indicated by their presence. Some of the insights contained in the book *The Spell of the Sensuous: Perception and Language in a More-than-Human World* by David Abram highlight reasons why we should take the Jain worldview seriously as one avenue of exploration for the development of an effective ecological outlook. Jainism proclaims that even seemingly inanimate things or objects, such as rocks and rivers, are in fact subjects possessing life force, or *jīva*, suffused with consciousness, energy, and bliss, as well as with a sense of touch. Although maintaining the primacy of the human perspective, David Abram suggests that we need to revive our relationship with things. Citing the French philosopher Maurice Merleau-Ponty, he writes that

> Our most immediate experience of things is necessarily an experience of reciprocal encounter—of tension, communication, and commingling . . . [W]e know the thing as a dynamic presence that confronts us and draws us into relation. . . . To define another being as an inert or passive object is to deny its ability to actively engage us and to provoke our senses. . . . By linguistically defining the surrounding world as a determinate set of objects, we cut our conscious, speaking selves off from the spontaneous life of our sensing bodies. . . . Only by affirming the animateness of perceived things do we allow our words to emerge directly from the depths of our ongoing reciprocity with the world.[24]

To illustrate this, Abram describes the experience of the forest:

> Walking in a forest, we peer into its green and shadowed depths, listening to the silence of the leaves, tasting the cool and fragrant air. Yet such is the transivity of perception, the reversibility of the flesh, that we may suddenly feel that the trees are looking at us—we feel ourselves exposed, watched, observed from all sides. If we dwell in this forest for many months, or years, then our experience may shift yet again—we may come to feel that we are a part of this forest, consanguineous with it, and that our experience of the forest is nothing other than the forest experiencing itself.[25]

In the words of Merleau-Ponty, "the presence of the world is precisely the presence of its flesh to my flesh."[26] Abram goes on to write that

> To the sensing body all phenomena are animate, actively soliciting the participation of our senses. . . . Things disclose themselves to our immediate perception as vectors, as styles of unfolding—not as finished chunks of matter given once and for all, but as dynamic ways of engaging the body and modulating the body. Each thing, each phenomenon,

> has the power to reach us and to influence us. Every phenomenon, in other words, is potentially expressive. . . . Thus, at the most primordial level of sensuous, bodily experience, we find ourselves in an expressive, gesturing landscape, in a world that speaks.[27]

Though this does not take the step of claiming that things possess the capacity to touch and feel, it does call for a greater acknowledgment of the power of the things of the world to shape our own perceptions and feelings.

In an earlier study, I explored a comparative analysis between Gaia theory and the Jain theory of the all-pervasiveness of eternal Jīva.[28] David Abram, alluding to Gaia theory, similarly suggests that the living-ness of things as articulated by Merleau-Ponty in fact has a scientific basis:

> We have at least come to realize that neither the soils, the oceans, nor the atmosphere can be comprehended without taking into account the participation of innumerable organisms, from the lichens that crumble rocks, and the bacterial entities that decompose organic detritus, to all the respiring plants and animals exchanging vital gases with the air. The notion of earthly nature as a densely interconnected organic network—a 'biospheric web' wherein each entity draws its specific character from its relations direct and indirect, to all the others—has today become commonplace. . . .[29]

Whether seen as a continuity of interchangeable life-forms or as a succession of discrete incarnations, the web-like nature of both contemporary biology and traditional Jain cosmology merits our attention. Both views require us to regard the world as a living, breathing, sensuous reality, from its elemental building blocks of earth, water, fire, and air, through its microbial expressions, right up to its array of complex insects and mammals, including primates. In the Jain tradition, this has led to a careful observance of the principle of nonviolence. In the world of contemporary ethics, it has led to the introduction of animal rights language, the argument for legal standing for trees, and most recently, the Great Ape Project, which advocates that full rights be accorded to chimpanzees, gorillas, and other high-functioning primates.

In contemporary forms of post-Christian spirituality in America, this has led to the emergence of reflection on the landscape as a means of attaining a heightened sense of intimacy, belonging, and meaning. This tradition, celebrated in the new anthologies of nature writing, has been part of American literature for over a century, as found in the writings of Henry David Thoreau, John Muir, Annie Dillard, Barry Lopez, and others. Abram's articulation of the appeal of this "practice" leads us to further explore those aspects of Jain philosophy that lend themselves to the valuing of particularity over a sense of oneness. Abram writes:

> There is an intimate reciprocity to the senses; as we touch the bark of a tree, we feel the tree touching us; as we lend our ears to the local sounds and ally our nose to the seasonal scents, the terrain gradually tunes us in turn. The senses, that is, are the primary way that earth has of informing our thought and of guiding our actions. Huge centralized programs, global initiatives, and the 'top down' solutions will never suffice to restore and protect the health of the animate earth. For it is only at the scale of our direct, sensory interactions with the land around us that we can appropriately notice and respond to the immediate needs of the living world. Yet at the scale of our sensing bodies the earth is astonishingly, irreducibly diverse. It discloses itself to our senses not as a uniform planet inviting global principles and generalizations, but as this forested realm embraced by water, or a windswept prairie, or a desert silence. We can know the needs of any particular region only by participating in its specificity—by becoming familiar with its cycles and styles, awake and attentive to its other inhabitants."[30]

One might object that this has nothing to do with Jainism, that the Jains do not wax eloquent about the landscape, that, at best, Merleau-Ponty and David Abram romanticize an unattainable weak monism, that the grim rigor of Jain ascetism must not be confused with transcendentalist elegy. But I would like to cite one compelling passage from the *Ācārāṅga Sūtra* and tell one story of my visit to Ladnun that, while not sentimental, nonetheless underscore the importance of sensory awareness in the Jain tradition.

Senses and Sensibilities in Jainism

In the second part of the *Ācārāṅga Sūtra*, Mahāvīra addresses his monks and nuns on the topic of forest preservation. This brief meditative advice encapsulates what could be seen as a textual foundation for the development of an activist Jain environmentalism. It also shows the timelessness of human greed and exploitation of the natural world. Mahāvīra tells the monks and nuns to "change their minds" about looking at big trees. He says, rather than seeing big trees as "fit for palaces, gates, houses, benches . . . , boats, buckets, stools, trays, ploughs, machines, [wheels] . . . , seats, beds, cars, sheds," they should speak of trees as "noble, high and round, big," with "many branches . . . , magnificent."[31] This indicates that Mahāvīra in fact did regard trees as inherently valuable for their beauty, strength, and magnificence and that he advised his followers to turn their thoughts from materiality by reflecting on the greater beauty of sparing a tree from the woodsman's ax.

While visiting Jain Visva Bharati in Ladnun, Rajasthan, I, by accident, stood atop an anthill of large red ants who swiftly moved up my pantleg. Had I been attuned to the local landscape, I would not have stood in such an inappropriate spot. I am an ant-sensitive person and quite adept at avoiding even the nearly microscopic lines of ants that parade along the sidewalks of Southern California. My Jain companions gently urged me to take care not to hurt the ants as I moved them back to the ground. This episode provoked in me multiple reflections: the Terāpanthī Śvetāmbara Jain community had located their combination monastery-seminary-university in the remote desert because few life-forms flourish there, thus reducing possible inadvertant harm. The Jains who visit this landscape frequently know how to scan the landscape (in this instance the ground under our feet) to avoid red ants; this reveals an intimacy with place. My companions felt compassion both for me in my error and for the ants as I somewhat awkwardly returned them to the ground. The entire scenario was filled with a heightened sense of immediacy and importance, a sort of meditation on the present in this simple encounter with ants.

Conclusion

Thomas Berry and Brian Swimme propose a new story based on scientific explanations (or "best guesses") regarding the origin and nature of the universe. In part, this approach depends on a starting point (the Flaring Forth or Big Bang) and the idea of an implied if not explicit sense of teleology. The Jain story does not include a fixed origin point in either assumed fact or metaphor, but rather, assumes the eternality of the world. It will not work as a story in the sense of a beginning, middle, and probable end. Rather, this system seeks to sacralize all aspects of worldly existence. By seeing all that surrounds us as suffused with life and worthy of worship, Jainism offers a different sort of story, a story that decentralizes and universalizes ethics, taking away overly anthropocentric concerns, and brings into vivid relief the urgency of life in its various elemental, vegetative, and animal forms. The cosmic story in the Jain tradition might well be a story of immediacy and care rather than of mythic structures and externally imposed ethical values.

At first glance, the Jain tradition might seem to be inherently ecologically friendly. It emphasizes nonviolence. It values all forms of life. It requires its adherents to engage only in certain types of livelihood, presumably based on the principle of *ahiṃsā*. Jainism's earth-friendly attitudes have been celebrated in L. M. Singhvi's *Jain Declaration on Nature* (reprinted in the appendix to this book), in Michael Tobias's video *Ahimsa* and its companion volume, *Life Force*, in my own book *Nonviolence to Animals, Earth, and Self in Asian Traditions*, and in the proceedings of the Ladnun conference on ecology and Jainism, the periodical *Jain Spirit: Advancing Jainism into the Future,* as well as in other materials. However, if we look at both the ultimate intention of the Jain faith as well as the actual consequences of some Jain businesses, we might detect a need for the sort of in-depth critical analysis and reflection that Thomas Berry has suggested for the Western world. For instance, Jains have long avoided using animal products in their many businesses. Lists of "green-friendly" materials could be developed by Jains to be used in manufacturing processes. The Jain programs of environmental education could be expanded to prepare future leaders to be more familiar with environmental issues. Jains could actively support air pollution reduction initiatives by making certain that their own automobiles in India conform to legal standards.

In some respects, however, environmental activism at best could earn a secondary place in the practice of the Jain faith. The observance of *ahiṃsā* must be regarded as ancillary to the goal of final liberation, or *kevala*. Ultimate meaning is not found in the perfection of nonviolent (in this case, eco-friendly) behavior but in the extirpation of all fettering *karma*. Although the resultant lifestyle for monks and nuns resembles or approximates an environmentally friendly ideal, its pursuit focuses on personal, spiritual advancement, not on a holistic vision of the interrelatedness of life. In terms of the lifestyle of the Jain layperson, certain practices such as vegetarianism, periodic fasting, and eschewal of militarism might be seen as eco-friendly. However, some professions adopted by the Jains, due to their religious commitment to harm only one-sensed beings, might in fact be environmentally disastrous, such as strip-mining for granite or marble, unless habitat restoration accompanies the mining process. Likewise, how many Jain industries contribute to air pollution or forest destruction or result in water pollution? The development of a Jain ecological business ethic would require extensive reflection and restructuring.

As Thomas Berry has noted, the task of ecological repair requires the networking of the political, economic, business, educational, scientific, as well as the religious communities. Jainism, given its ethic of nonviolence and its deep involvement with the governmental structures of India and the business community worldwide, is well equipped to initiate the process. But, in order for any of this work to be effective, it must proceed from a story. The story of the human superiority over nature has been told throughout the world, even by the Jains who seek to rise above nature. And this story has been realized, as seen in the success of consumer culture worldwide. Native habitats continue to be destroyed as industrialization expands. As this happens, entire species of animals, insects, and plants disappear, never to return. Yet humans proliferate, taking up more space worldwide with their houses and condominiums and farmland, encroaching on and destroying the wild, isolating humans within fabricated landscapes that separate the human from the pulse of nonhuman life.

A shift in consciousness must take place that places greater value on life in its myriad forms. The cosmological views of Jainism, the insights of contemporary science, and the growing perception of the beauty and fragility of the natural order all can contribute to this essential shift toward the development and enhancement of an earth-friendly way of life.

Notes

1. Śānti Sūri, *Jīva Vicāra Prakaraṇam along with Pāthaka Ratnākara's Commentary*, ed. Muni Ratna-Prabha Vijaya, trans. Jayant P. Thaker (Madras: Jain Mission Society, 1950), 163. Hereafter cited as *JVP*.

2. Umāsvāti, *Tattvārtha Sūtra; That Which Is*, trans. Nathmal Tatia (San Francisco: HarperCollins, 1994), 2.2.333; p. 53.

3. Ibid., 2:24; pp. 45–46.

4. Brian Swimme, *The Hidden Heart of the Cosmos: Humanity and the New Story* (Maryknoll, N.Y.: Orbis Books, 1996), 93.

5. Ibid., 33.

6. Ibid., 112.

7. See Brian Swimme and Thomas Berry, *The Universe Story: From the Primordial Flaring Forth to the Ecozoic Era: A Celebration of the Unfolding of the Cosmos* (San Francisco: HarperSanFrancisco, 1992).

8. *Ācārāṅga Sūtra* 1.8.1.11–12. From *Jaina Sutras*, Part 1, *The Ākārāṅga Sūtra. The Kalpa Sūtra*, trans. Hermann Jacobi (1884; reprint, New York: Dover, 1968). Hereafter cited as *AS*.

9. See R. Williams, *Jaina Yoga: A Survey of the Mediaeval Śrāvakācāras* (London: Oxford University Press, 1963).

10. *AS* 2.3.3.3.

11. *JVP* 34.

12. *JVP* 163.

13. *JVP* 3–4.

14. *JVP* 5.

15. *JVP* 6.

16. *JVP* 7.

17. *JVP* 15.

18. *JVP* 16, 17.

19. *JVP* 35.

20. *JVP* 18.

21. Thomas Berry, *The Dream of the Earth* (San Francisco: Sierra Club Books, 1988), 164.

22. Thomas Berry, "Religion in the Ecozoic Era" (plenary address to the annual meeting of the American Academy of Religion, Washington, D.C., November 1993), 2.

23. Ibid., 18.

24. David Abram, *The Spell of the Sensuous: Perception and Language in a More-than-Human World* (New York: Pantheon Books, 1996), 56.

25. Ibid., 68.

26. Maurice Merleau-Ponty, *The Visible and the Invisible*, trans. Alphonso Lingis (Evanston, Ill.: Northwestern University Press, 1968), 127.

27. Abram, *The Spell of the Sensuous*, 81.

28. Christopher Key Chapple, *Nonviolence to Animals, Earth, and Self in Asian Traditions* (Albany: State University of New York Press, 1993), chap. 4.

29. Abram, *The Spell of the Sensuous*, 85.

30. Ibid., 268.

31. *AS* 2.4.2.11–12.

Ecology, Economics, and Development in Jainism

PADMANABH S. JAINI

Fundamental Jain Teachings

While the Jains are undoubtedly adherents of one of the most ancient religious traditions in the modern world, they constitute one of the smallest groups, only slightly larger than the Zoroastrians. According to the latest government census, Jains number less than six or seven million people, or less than one percent of the entire Indian population. Even though the size of the Jain community has never equaled that of other religious groups, it has been highly influential throughout Indian history because of its heavy concentration in commerce and industry. For this reason, most Jains have tended to live in large urban areas, where they have remained in close contact with the governing powers, whether Rajput, Mogul, the British East India Company, or the government of India after independence.

One of the distinguishing features of Jainism is that there is no belief in a creator god (*īśvara*). Hence, Jains do not believe that everything in the world, including plants and animals, was created by an intelligent first-cause for humankind's benefit and consumption. However, Jains do believe in the attainment of enlightenment and salvation from the repeated cycle of death and rebirth by their founding teachers, called Jinas (Spiritual Victors), from whom the name Jain is derived, and the possibility for their followers of the same attainment.

This is achieved not through the grace of a deity but via one's own exertion and personal dedication to the path of spiritual purification. This path involves mental practices of meditation, physical practices of self-denial and austerities, and avoiding harm (*ahiṃsā*) to all living beings. Jains believe that these practices are necessary for an individual embodied soul to rid itself of the effects of *karma*—the accumulated results of actions performed over many lifetimes—and to minimize the accumulation of new *karma*. These actions proceed from the soul's passions for nourishment (*āhāra*), reproduction (*maithuna*), and the accumulation of worldly goods (*parigraha*) for the attainment of power over others. The salvation of the soul lies in its inherent ability to overcome these beginnningless passions through knowledge of the true nature of the self—a self capable of being free from embodiment (*mokṣa*) and attaining the resultant infinite happiness (*sukha*) associated with that freedom.

In order to progress along the path to salvation, Jains believe that it is necessary to reduce to a minimum actions that result in harming other living beings and attachment to and accumulation of excessive personal possessions. In the words of the twenty-fourth Jina of our world-age, Mahāvīra: "No being in the world is to be harmed by a spiritually inclined person, whether knowingly or unknowingly, for all beings desire to live and no being wishes to die. A true Jain, therefore, consciously refrains from harming any being, however small."[1]

The exemplars of these values are members of the Jain mendicant community, those men and women who have renounced the household life and are totally dedicated to the path of salvation, which is characterized by the practice of complete nonviolence (*ahiṃsā*), adherence to truthfulness (*satya*), not taking anything that is not given (*asteya*), lifelong celibacy (*brahmacarya*), and nonpossession (*aparigraha*), except for those items requisite for a mendicant life. Thus, they live a life of economic poverty, consuming the minimum amount of food and material goods necessary to sustain life. However, this poverty is voluntarily assumed and in many cases is preceded by renouncing an affluent lifestyle with a ritual act of giving away considerable amounts of wealth. The number of Jains who have chosen to live such a difficult and austere life has always been small but substantial. There are at present about twenty-five hundred male mendicants (*sādhus*) and almost six thousand female mendicants (*sādhvīs*).[2]

Laypeople, the *śrāvaka*s and *śrāvikā*s (literally, listeners), not only maintain these ascetics with gifts of food and shelter but also strive to practice, to the best of their ability, the virtues of nonviolence and nonpossession within the limitations of a prosperous household life. The polarity of householder and ascetic is said to be one of the most characteristic features of the Jain community. Indeed, "a creed of complete otherworldliness has offered a background for the successfully worldly."[3]

Jain Involvement in Worldly Activities

Prosperity in the world, in the Jain view, is a legitimate goal for a layperson in pursuit of the higher goal of renunciation, because it affords the opportunity not only for earning a proper livelihood but also for offering gifts (*dāna*) to those who follow the path of renunciation and for extending charity and service (*sevā*) to those who are in need. Prosperity is invariably joined with the ideas of purity in the means of acquisition, integrity in the means of its investment, liberality in its distribution, and restraint in its enjoyment. Adopting a legitimate means of livelihood (*nyāyopatta-dhana*) is extremely important for a Jain, since the chosen occupation determines the degree to which violence can be restricted.

As early as the sixth century, manuals were composed for the laity that contained long lists of occupations that were considered to be unsuitable for a practicing Jain. These unacceptable livelihoods derived from 1) the production and sale of charcoal; 2) the destruction, including the sale, of timber; 3) construction and sale of carts; 4) the use or rental of one's own animals for transporting goods; 5) animal by-products, such as ivory, bones, conch-shells, pelts, down; 7) trade in lac and similar substances; 8) trade in alcohol and forbidden goods; 9) trade in slaves and animals; 10) trade in destructive articles, such as weapons, farming implements, poisons; 11) the operations of mills and presses that crush sugar cane and seeds; 12) livelihood from the mutilation of animals, including gelding and branding; 13) work that involves the use of fire, such as clearing forests or meadows for cultivation; 14) work that involves draining lakes for future cultivation; and 15) work that involves breeding or rearing destructive animals, such as monkeys and so forth.[4] The rationale for such a list is based on the belief that souls exist in animals and plants; thus, these life-forms are to be respected and not harmed. All living beings have the capacity to experience pleasure and pain and have the wish to live, and some souls eventually may take birth as human beings. Therefore, Jains have preferred whenever possible to engage in those occupations where harm to human beings, animals, and plants is minimized. However, when injury does arise in the course of performing one's occupation (*ārambhaja-hiṃsā*), it is thought to produce less severe negative karmic effects than acts of violence promoted by greed and anger.[5]

Lists of permissible occupations include those associated with the merchant community, such as trade in precious stones, banking, commerce, clerical activities; the practice of traditional forms of medicine; arts and crafts; and service to the government or ruler. Records indicate, for example, that many Jains living in Patan, the old capital city of Gujarat, during the nineteenth century were shopkeepers, moneylenders, pawnbrokers, and landlords of agricultural lands in the villages nearby. Textile manufacture was an important business for Jains in Ahmedabad. With the growth of manufacturing industries, Jains entered the hardware, machinery, chemical, and pharmaceutical businesses.[6] Like their counterparts in earlier times, when Jain businessmen investigate business opportunities for investing their capital today, they weigh not only potential profits and losses but also the degree to which *hiṃsā* is inherent in each occupation. It has been reported, for example, that a group of Jains in India rejected lucrative investments in luxury hotels, mostly catering to Western tourists, because the hotels had to serve nonvegetarian food. Instead, they chose to invest in the newspaper industry, namely, in the Times of India Group, which additionally helped to educate the nation.[7]

In order to increase profits, Jains would not have engaged in the practice followed by certain agriculturists of serving unnatural foods to animals, such as the bonemeal and animal by-products fed to cattle that then engendered the recent tragic outbreak of mad cow disease and the subsequent slaughter of millions of innocent animals. It goes without saying that Jains, as strict vegetarians and living in accordance with their doctrine of *ahiṃsā*, would not have raised cattle for any purpose whatsoever except for supplying dairy products. Given the importance of agriculture, it too is included in the list of acceptable occupations, although, if possible, Jains themselves do not actually till the land.

Military service is a permissible occupation for a Jain layman, under certain circumstances, namely, as a last resort in guarding the interests of one's prosperity, honor, family, community, or nation. Thus, Jains are not total pacifists: if all efforts at a peaceful resolution to a conflict fail, a layman may counter violence with violence. However, what distinguishes the Jain approach even to a "just war" from that of several other world faiths is that while valor on the battlefield is preferable to cowardice, these acts of bravery do not lead to rebirth in heaven unless a warrior renounces all possessions and becomes a mendicant.[8]

One of the factors that has enabled Jain businessmen to prosper has been the strong bonds of the family and the support of other members of the community. Family members have worked together to make their businesses prosper and, when financial assistance was needed, it was forthcoming from other Jain businessmen. Thus, this minority community has been able to depend on itself for economic growth and survival.[9]

The benefits of acquisition of wealth by legitimate means on the part of the lay community has long been recognized within Jainism. In the eighth-century *Dharmabindu* of Haribhadrasūri, there is a description of the qualities of an ideal layman (*śrāvaka-guṇas*).[10] The first and foremost of these qualities is that of being endowed with honestly earned wealth (*nyāya-sampanna-vibhava*), for wealth acquired by honest means not only brings freedom from anxiety in this world, it also leads to a happy reincarnation. Moral integrity in business is a necessary element in the long-term acquisition of wealth. Economic prosperity gained by dishonest means is transitory in nature. According to Ācārya Hemacandra (1088–1172), the author of the monumental *Jain Yogaśāstra*, honestly earned wealth is that which has not been acquired by treason, betrayal of friends, breach of trust, theft, false witness, false weights and measures, or deceitful speech.[11] In our own times, the Anuvrata Movement, started in 1949 by a Jain mendicant, the late Ācārya Tulsi (1914–1997), is founded on a commitment to observing ethical standards and restraint in all aspects of life. The code of conduct for this movement includes the observance of moral integrity, especially in business.[12]

Wealth, Donations, and Spiritual Service

A related concept is the accumulation of wealth for the spiritual benefits (*puṇya*) gained by its diligent sharing with the ascetic, the righteous, and the needy. This entails offering, with veneration, freshly cooked vegetarian food, drink, shelter, and other requisites for life to mendicants, the most worthy to receive such gifts, as well as offerings out of compassion to those who are poor or afflicted. Hemacandra defines a great layman (*mahāśrāvaka*) as "one who strews wealth out of devotion in the seven fields (*kṣetra*) and out of compassion (*dayā*) on the needy." These seven fields, or those who are worthy recipients, in order of importance, are: images of the Jina (*Jina-bimba*), Jain temples (*Jina-bhavana*), Jain religious texts (*Jināgama*), male Jain mendicants (*sādhus*), female Jain mendicants (*sādhvīs*), Jain laymen (*śrāvakas*), and Jain laywomen (*śrāvikās*).[13] By meritorious acts of giving (*dāna*), *karma* leading to well-being in this life and in lives to come is accrued, while negative *karma* is destroyed. Since mendicants cannot be involved in performing tasks necessary for survival in the world, such as lighting a fire or cooking food, the daily activities performed by members of the Jain lay community are essential for the survival of the entire Jain community. "Attaining physical wellbeing, earning wealth, and spending that wealth towards the greater glory of the Jain religion are therefore not just necessary, they are also laudable."[14] In this manner, an appropriate balance between the three legitimate goals of a layperson, namely, religious pursuits (*dharma*), earning a livelihood (*artha*), and maintenance of the family lineage (*kāma*), is maintained. With the legitimate means of earning one's livelihood thus secured in accordance with the guiding principles of nonviolence and honesty, and with an understanding of the importance of sharing one's wealth with others, Jains have been able to achieve for themselves a measure of wealth and prosperity, while at the same time contributing to the well-being of society at large.

Distribution of wealth, however, is not confined to the worthy recipients mentioned above. Charity and service (*sevā*) is extended to those of other faiths who are in need, and protection and care are extended to animals as well. In the words of Hemacandra, "a [great layman] should use his wealth indiscriminately to assist all who are in misery or poverty, or who are blind, deaf, crippled, or sick, irrespective of whether [ideal] recipients or not."[15] The Jain community has been at the forefront in providing fresh vegetarian food and clean water during times of drought or natural disaster. For example, at the time of the recent devastating earthquake in Maharashtra, the Jain community established soup kitchens to feed thousands of people who were displaced from their homes. Providing food and water for animals is another service performed by Jains. Through the building of shelters for animals, including birds, and by providing food and medicine for animals at these shelters, Jains perform the services of giving protection to (*abhaya-dāna*) and showing compassion for (*jīva-dayā*) all living beings.[16] It has been estimated that during the famine in Gujarat in 1988 Jains supported many of the state's voluntary relief agencies and, in addition, donated hundreds of thousands of rupees to rescue animals from the effects of this drought.[17] Within the communities in India where meat-eating is practiced, Jains have lobbied from ancient times for a prohibition of the slaughter of animals (*amāri*) on certain holy days.

Conservation of plant life also is demonstrated in the practice of preparing only the amount of food that can be consumed without waste or spoilage. Concern for insects is shown by cleanliness in areas where food is prepared and consumed so as not to attract insects that might inadvertently be killed, as well as in the traditional practice of not eating after dark and not consuming honey. Protection of even the least developed forms of life is seen in the straining of drinking water, which is then boiled in order to purify it, and in not eating figs and other such fruits that are thought to contain minute life-forms.[18] Providing medicines and medical services (*aushadha-dāna*) is demonstrated by the establishment of and donations to hospitals and medical clinics, such as the eye and orthopedic clinics operated by Veerayatan in Bihar. In recent times, Jain physicians have volunteered their services to perform reconstructive surgery, eye surgery, and so forth.

The importance of knowledge and education within Jainism is shown in gifts of knowledge (*jñāna-dāna*) through the establishment of schools and donations to scholarship funds, the publishing of texts, and the maintenance of manuscript libraries at temple complexes (*bhaṇḍāras*), which house Jain as well as Hindu and Buddhist manuscripts or rare printed works. Education and literacy for both men and women has been a priority for the Jain community because education enables one to read religious texts and ultimately gain knowledge of the soul, as well as properly conduct household and business affairs. The importance of proper knowledge is reflected in the expression, "First knowledge, then compassion" (*paḍhamaṃ nāṇaṃ tao dayā*).[19]

The Social Role of Jain Women

While, historically, men have engaged in family occupations and thus have been responsible for generating income, women have played a significant role in the accumulation and maintenance of wealth as well as in its distribution to others. Frugality in managing the household is demonstrated by the non-wasteful preparation of fresh food that the family consumes each day and that is offered to Jain mendicants who visit the household on their begging rounds. The ability to give food to mendicants is a great honor for a family, not only because of the religious merit accrued through such an act, but also because it is an indication that the food was prepared with the utmost attention to its purity and that the women who prepared it strictly adhere to Jain dietary restrictions. Fasting is an important part of the religious practices of many Jain women, who regularly restrict the type and quantity of food consumed or refrain from eating altogether for a time. Transmission of family and religious values to children has been primarily the responsibility of women, who have had more time to devote to religious activities (*pūjā*) in the home or at the temple.

Women in the Jain community have had the freedom to choose between the household life and a life of mendicancy. According to Jain chronicles, there has always been approximately twice as many female mendicants as male.[20] It is common for women to assume roles of leadership in the mendicant community by offering religious discourses and providing religious services to the laity. Therefore, education of women has been a priority of the Jain community, and their literacy is part of the Jain heritage.

Women also have had some degree of economic freedom, not just in managing the household but also in their ability to donate money in their own names. As early as the second century A.D. many inscriptions are found at Mathura with the names of the women who sponsored the consecration of images.[21] Indeed, in these records, the names of women vastly outnumber those of men. Tamil inscriptions from the eighth to twelfth centuries also provide evidence for the ongoing role of women in religious giving.[22]

Jainism, Ecology, and Social Welfare

Because of the emphasis placed on nonharm (*ahiṃsā*) of all living beings, ecology is increasingly becoming a focus of the Jain community. Recognizing the importance of this endeavor, representatives of the Jain community presented a Declaration on Nature on 23 October 1990 at Buckingham Palace, thus joining the World Wide Fund for Nature (WWF) Network on Conservation and Religion. This declaration has been put into practice through the establishment of the Ahimsa Environmental Award, which encourages individuals, businesses, and community groups to practice Jain principles in environmental ethics. Criteria for this award include ways that the organization or group has reduced its production of harmful waste; how open the organization is about the impact of its practices on the environment; the extent to which it makes use of sustainable and nondepleting resources; the extent to which it has reduced its use of natural resources; and the ways in which it implements the sharing of the earth's resources.[23]

The Jain community has also presented a Statement on Ecology and Faith at the Alliance of Religions and Conservation (ARC). Jains were represented at the 1992 UN Conference on Environment and Development ("Earth Summit") at Rio de Janeiro. In an effort to meet the environmental challenges that face this world as it continues on the path of economic development and industrialization, Jains want to strengthen links with other faiths and with organizations to form a strong lobby on matters of ecology. They would like to adopt a program to raise public awareness of the responsibility of individuals, groups, and governments for the care of nature and the preservation of life of all forms, and they will seek the systematic introduction of the environment as a subject of study in schools and universities.[24] The Jain Vishva Bharati Institute, which was founded in 1970 in Ladnun, Rajasthan, by the late Ācārya Tulsi, offers a program in Ecology and Environmental Science. The Third International Conference on Peace and Nonviolent Action, which was held there in 1995, was attended by delegates from twenty-one countries. The objective of this conference was to develop a global plan for the protection of the earth and its inhabitants and to emphasize the necessity for a simple lifestyle based on *ahiṃsā* and spirituality.[25]

In accordance with the guiding principles of their religion, the Jain community is undertaking projects aimed at addressing social and environmental problems. Recognizing that the restoration of Jain pilgrimage sites incorporates spiritual benefits and conservation, the community has initiated a series of comprehensive forestation and educational projects at their largest pilgrimage sites. The Greening of Palitana Project at the pilgrimage center of Shatrunjaya (Śatruñjaya) in Gujarat, which is being underwritten by several prominent Jain Trusts, involves planting trees on the hills, the approaches to the pilgrimage center, and at the various facilities for the pilgrims. There are efforts to include an active and visible ecological awareness component so that the pilgrims will be motivated to take back with them the environmental message as an ethical mandate of their faith. As this project is implemented, similar projects will be started at other pilgrimage centers sacred to the Jains, such as Shravanabelgola in Karnataka.

The Akhil Navsarjan Rural Development Foundation—Bombay (ANARDE Foundation) has undertaken massive forestation projects and the development of wastelands in India and promotes rural development through programs that assist people to help themselves. The programs are divided into six categories: economic, medical, education, social, cultural, environmental. Working through eighteen rural development centers and the Forest Department's agricultural and forest universities, the ANARDE Foundation has developed several plant nurseries with the help of schools, farmers, village communities, and women's groups.

Jainism and Development: Contemporary Examples

In accordance with the Jain belief that nonharming is the highest religious tenet (*ahiṃsā paramo dharmah*) and that *ahiṃsā* necessitates placing limits on possessions (*aparigraha*), criteria for development should be guided by the principle of least harm and should encourage conservation of resources. In the words of Mahatma Gandhi, "Strictly speaking, no activity is possible without a certain amount of violence, no matter how little. Even the process of living is impossible without a certain amount of violence. What we need to do is to minimize it to the greatest extent possible."[26]

Unfortunately, rapid and often unregulated industrial development and a huge increase in population have resulted in harmful air pollution, global warming, contamination of water and earth, urban migration, and the disintegration of traditional family structures. While these negative effects could be curtailed by eliminating all new development and educating people to be satisfied with consuming only the minimum necessary for survival, this approach would be extreme and take into account only one perspective. Jains strongly adhere to a belief that reality has many facets and that views must be "not-one-sided" (*anekāntavāda*). The potential benefits that could accrue from a proposed development project to alleviate suffering and improve the quality of life of one segment of society must be weighed against its negative impact on other humans, as well as on animals, plants, earth, water, and air. In establishing a portfolio of development projects, there should be a balance between funding for projects that use currently available technologies and new technologies for future use. The short-term economic need for income should be balanced against the social and emotional needs of individuals to live in a clean, safe, and nurturing environment. In examining how a balanced approach to development could be implemented, two projects in India from the energy and health care sectors recently funded by the World Bank may be examined from a Jain perspective.

In September 1997, the World Bank announced the funding of the Coal Sector Rehabilitation Project and Coal India, Ltd., in support of the Government of India's position that coal-based thermal power generation will be the key element in India's energy development projects. This project involves opencast mining operations, which affect large areas of land and sometimes necessitate the resettlement of entire communities.[27] How should such a project be viewed? In the short-term, in order to continue development, India needs to be assured of a sustained supply of electricity, and coal is often the fuel of choice. While this method of energy production is preferable to hydroelectric power generated by large-scale damming projects, such as the Narmada River project, it still is destructive to the ecosystem. The World Bank must ensure that the negative impact is addressed and the affected people rehabilitated.

In evaluating projects that involve the construction of new power plants that burn this coal, an indicator that could be used is the degree to which harm to the environment is minimized by incorporating the most advanced technology. For example, the Export-Import Bank of the United States is financing a power plant in Gujarat that uses integrated gasification combined cycle (IGCC) technology, which produces a clean-burning gas from low-grade coal, reducing the negative impact on the environment and allowing India to use less expensive local resources.[28] This coal project should be balanced by funding for nonpolluting and renewable energy sources, such as wind, solar, biogas, and small-scale hydro, which are being promoted by the Ministry of Non-conventional Energy Sources, that can be used for fueling homes now and could supply energy on a larger scale with technological advances in the future.[29] The funding of development for energy production should be balanced by projects that reduce the need for energy consumption in the first place by using more energy efficient electric motors and utilizing more efficient methods of manufacturing.

Another project that can be evaluated using this balanced approach is the malaria control project.[30] Controlling the spread of malaria has, in the past, involved the killing of millions of mosquitoes by spraying insecticides that are harmful to the environment. Although this project still involves killing, there is an effort to reduce harm to insects and the environment by emphasizing multiple means of intervention with a decrease in the use of insecticides, more focused spraying with less environmentally harmful chemicals, and the increased use of medicated mosquito nets. The degree to which killing and environmental harm is reduced in comparison to past projects could be an indicator in evaluating a project such as this.

Conclusion

Development, by its own nature, must be limited to some degree. According to Ācārya Mahāprajña: "It becomes imperative for us to limit the concept and dimension of development aroused by greediness. We are of the opinion that in a state of unrestrained passions and unbalanced development the very conception of living in harmony with nature is distorted. . . . [We] should develop a balanced lifestyle by controlling economic development and consumption."[31] The aim of development, particularly that which is funded by loans of such large amounts, is to bring prosperity to a community of people who have suffered age-long poverty. The Jains, with their approach to a balanced view of life, have managed throughout history to maintain a reasonably high standard of living while at the same time helping to raise the standard of life, both material and spiritual, for non-Jains.

For Jains, prosperity is the ability to be self-reliant and to contribute to the well-being of one's family by providing for its material needs, such as food, shelter, education, and medical care, as well as its spiritual needs. It is the ability to accumulate wealth that can be given to those who are reliant on others for basic needs, whether this reliance is the result of voluntary poverty circumstances or natural disasters. However, any economic prosperity will be short lived if the social structure inherent in the family and community that can provide a foundation for inculcating moral and spiritual values is not maintained. Those who are impoverished lack shelter and sufficient food to avoid malnutrition. Acts of violence and greed must in the long run result in poverty and pain, whereas acts performed out of generosity, charity, and compassion lead to a happier and more prosperous life. In the words of Ācārya Mahāprajña:

> Lifestyle embedded entirely in materialistic consideration doesn't provide a basis for living in harmony with nature. And a lifestyle based exclusively on a spiritual foundation does not prove adequate for the journey of life. In order to make our lifestyle complete in every sense we need a new outlook on life. A harmonious method of material, economic, and spiritual development can be devised on its basis. Such method can never ignore the balance between intellectual and emotional needs.[32]

The Jain response to development issues must be mindful of traditional Jain teachings on nonviolence and nonpossessiveness. According to the Jain perspective, development decisions need to follow the criteria of least harm, conservation of resources, and the degree to which a mix of projects represents a balance between destruction and preservation of the environment.

When making decisions about development and ecology, Jains seek to strike a balance between the spiritual needs and the material prosperity of their own community. Additionally, they are concerned with the well-being of the world surrounding them, including the animal and vegetable kingdoms. Although the lay community is called upon to engage in economic development, such activities are to be pursued within the guidelines of minimal harm to all living beings (*ahiṃsā*) and moderation in the enjoyment of wealth for personal gratification (*aparigraha*). At the dawn of a new millennium, the Jain community draws its inspiration for a global perspective when addressing issues of prosperity and poverty from the ancient Jain words of wisdom, "All life is bound together by mutual support and interdependence" (*parasparopagraho jivānām*).[33]

Notes

This chapter was originally presented at the World Bank's World Religions meeting, Lambeth Palace, London, February 1998, and published in *World Faiths and Development* by the World Faiths Development Dialogue, 1998.

This chapter was written in consultation with L. M. Singhvi and Nemu Chandaria of the Institute of Jainology. I also wish to thank Kristi Wiley for her assistance in preparing this paper.

1. Padmanabh S. Jaini, *The Jaina Path of Purification* (Berkeley: University of California Press, 1979), 66.

2. Padmanabh S. Jaini, *Gender and Salvation: Jaina Debates on the Spiritual Liberation of Women* (Berkeley: University of California Press, 1991), 25.

3. R. Williams, *Jaina Yoga: A Survey of the Mediaeval Śrāvakācāras* (Oxford: Oxford University Press, 1963), xxii.

4. Ibid., 117–23.

5. Jaini, *The Jaina Path of Purification*, 171.

6. John E. Cort, "Liberation and Wellbeing: A Study of Śvetāmbara Mūrtipūjak Jains of North Gujarat" (Ph.D. diss., Harvard University, 1989), 83. A revised version of this work has since been published under the title *Jains in the World: Religious Values and Ideology in India* (Oxford: Oxford University Press, 2001).

7. Michael Tobias, *Ahimsa: Non-Violence* (PBS Film, Los Angeles, Direct Cinema, 1989).

8. See the story of Varuṇa in the *Bhagavatī Sūtra* (320–321), in Jozef Deleu, *Viyāhapannatti (Bhagavatī)* (Brugge: Rijksuniversity te Gent, 1970), 142.

9. For a discussion on the organization of the Jain business community, see Paul Dundas, *The Jains* (London: Routledge, 1992), 168–71; and Marcus Banks, *Organizing Jainism in India and England* (Oxford: Clarendon Press, 1992).

10. Williams, *Jaina Yoga*, 256–70.

11. Ibid., 260.

12. Dundas, *The Jains*, 223–24. *Young Jains International Newsletter* publishes articles on business and ethics and environmental issues. See, for example, October–December 1992 and October–December 1994.

13. Cort, "Liberation and Wellbeing," 243–44.

14. Ibid., 468–69.

15. Williams, *Jaina Yoga*, 165.

16. See D. O. Lodrick, *Sacred Cows, Sacred Places* (Berkeley: University of California Press, 1981).

17. *India Today* (International Edition), 15 July 1988, as cited in Dundas, *The Jains*, 170. For the rescue of animals, see *Young Jains International Newsletter*, October–December 1994.

18. Jaini, *The Jaina Path of Purification*, 167.

19. Ibid., 66.

20. Ibid., 37.

21. Heinrich Luders, *Appendix to Epigraphia Indica*, vol. 10, *A List of Brahmi Inscriptions* (Calcutta: n.p., 1912).

22. Leslie C. Orr, "Jain and Hindu 'Religious Women' in Early Medieval Tamil-

nadu," in *Open Boundaries: Jain Communities and Cultures in Indian History*, ed. John E. Cort (Albany: State University of New York Press, 1998).

23. *Ecology and Faith: Jainism*, World Wide Fund for Nature, 1995.

24. Ibid.

25. *Anuvibha Reporter* (Journal of Anuvrat Global Organization, Jaipur), 3, no. 1 (1997).

26. As quoted in Christopher Key Chapple, *Nonviolence to Animals, Earth, and Self in Asian Traditions* (Albany: State University of New York Press, 1993), 53.

27. World Bank Group, "World Bank $532 Million Loan to Fuel India's Vital Coal Industry," news release no. 98/1472/SASIA; http://www.worldbank.org.

28. Environment News Service, EnviroLink Network on the World Wide Web, 9 January 1997.

29. *India News* (Embassy of India, Washington, D.C.), December 1997, 4–6.

30. World Bank Group, "India Adopts Improved Approach to Malaria Control," news release no. 97/1385SAS; http://www.worldbank.org.

31. Keynote address delivered at the Third International Conference on Non-violent Action, published in *Anuvibha Reporter* (Journal of Anuvrat Global Organization, Jaipur), 3, no. 1 (1997): 57–58.

32. Ibid.

33. *Tattvārtha Sūtra* 5.21.

Voices within the Tradition: Jainism *Is* Ecological

The Environmental and Ecological Teachings of Tīrthaṅkara Mahāvīra

SADHVI SHILAPI

The Ecological Crisis

Industrial growth, technological advancement, population overload, over-exploitation of natural resources, and the brutality of the human community toward other life-forms on the planet have caused an unprecedented environmental and ecological crisis. If one analyzes all these causes, it is clear that at the center of this crisis lies the human race. It is the human community, among all the forms of life on this planet, that is interfering with the laws of nature by squandering her gifts or by destroying the existence of other species for personal and greedy reasons. Ironically, this issue of protecting the environment and the ecology has not come up because we have begun to hear the cries of plants, animals, and other forms of life—a direct result of our maltreatment of them—but because human life itself is in danger. One often hears such concerns and questions as Daniel Maguire's "If current trends continue, we will not," or Thomas Berry's "Is the human a viable species on an endangered planet?"

We worry about the thinning of the ozone layer and the poisoning of the air, the land, and the water, which comprise the basic elements for the substance of our life. The depletion of these elements may result in a legacy for our children and grandchildren of compromised immune systems, increased infectious disease and cancer rates, destroyed plants and the disruption of the food chain. I wonder, would we have the same concern, the same worry for other forms of life, if human life had not been in danger? Would we even have given the protection of the environment a second thought? It seems as if we are more concerned about the welfare of the human race, which sometimes conveys the impression that somehow, if we were able to protect human life and eradicate the danger surrounding it, we might again be able to forget the issue of the protection of other life-forms and not be concerned about their continued existence.

Mahāvīra's Life Ethics

Tīrthaṅkara Mahāvīra endowed humanity with a fundamental thought upon which behavior patterns and relationship with the environmental ideally could be based. An equality of all forms of life and reverence for all of them is his central teaching. He taught, "As you want to live, do so to others." In that definition of "others," he embraced not only all living beings that can move, but also the existence of earth, air, water, and vegetation. He even expressed a reverence for the inert. By making people aware of the existence of life in earth, air, water, and vegetation, he made a fundamental contribution to our understanding of ecology. He considered any injury to any of these a sinful act. The *Ācārāṅga Sūtra* (ca. 400 B.C.E.), the first main scripture of the Śvetāmbara Jain tradition, describes the thoughts of Mahāvīra on the life in vegetation: "Vegetation has life just as human beings have life. It is born as are human beings, its body grows and feels pain when pricked or cut with weapons. Like human beings, vegetation requires food. So consumption or use of vegetation in such a way may lead to his or her misfortune."[1] Mahāvīra further proclaimed that anyone who neglects or disregards the existence of earth, air, water, and vegetation disregards his own existence, which is intrinsically bound up with them.

He actively propagated these ideas and went on to make the protection and care of life in all its manifestations an obligatory duty for all Jains. For example, the daily prayer of the Jains contains a word of forgiveness for any harm or pain caused not only to fellow human beings but to all forms of life. A Jain devotee says:

> I confess to any injury caused by the path of my movement, in all my comings and goings, in treading on living things, in treading on seeds, in treading on green plants, in treading on dew, on beetles, on mould, on moist earth, and on cobwebs; whatever living organisms with one

> or two or three or four or five senses have been injured by me or knocked over or crushed or squashed or touched or mangled or hurt or frightened or removed from one place to another or deprived of life, I confess to that.[2]

However, this does not mean that Mahāvīra was not conscious of the fact that our very survival involves violence of one kind or another. In fact, his disciples even said to him, "According to your philosophy and teaching, whatever we do, whether it be eating, drinking, walking, sitting or even breathing, we are committing violent acts, interfering with nature, destroying the environment." Lord Mahāvīra said, "If you are aware of all your actions, and are careful about what you do in relation to other living things, you will develop spirituality and be in perfect harmony with the natural world."

Carefulness in all actions became the norm for the Jain way of life. This developed to the extent that those who renounce the world are required to take precautions for the protection of life before every action they perform. For instance, every morning and evening they should check all their belongings to ensure that no little creature has been hurt or killed, even unintentionally. Jain devotees are dissuaded from throwing any waste into river and lakes and are urged to guard the lives of all creatures found in these precious natural environments. Instead of just dumping waste anywhere, a place should be found where the least harm would be done to other living organisms.

All these examples show the deep concern that Tīrthaṅkara Mahāvīra had for ecology, which comprises a balance between plants, animals, and humans. If a human being is murdered by someone, the killer is punished by law; if animals are slaughtered or mistreated in any way, the perpetrator of the crime goes unpunished. This has become an accepted way of doings things in our society. If the entire human race were to adopt as a way of life a caring attitude toward animals, vegetation, or any other living thing, if we were to treat the breaking of this rule as a crime, the global environment scenario would change enormously.

Trees and Forests in Jain Literature

There are many stories and anecdotes in the commentaries to the Jain scriptures about the voices of plants and animals being heard by people. One of them goes like this. Long, long ago, a Brahmin had a majestic banyan tree. The shade of its widespread branches was cool and lovely and its shelter broadened out to encompass twelve leagues. No one needed to guard the fruit, and no one did any injury to another in picking the fruit. Then, one day a man came along who ate his fill of the fruit. He deliberately broke off a branch and then went on his way. The spirit who dwelt in the tree thought, "How amazing, how astonishing that anyone could, after eating his fill of fruit, be so evil as to break off a branch of this magnificent tree. What if the tree should just stop bearing any fruit?" It is also said that the spirit who resides in that tree came to Tīrthaṅkara Mahāvīra and complained that someone had cut off his child's arm. Lord Mahāvīra, protector of this seemingly mute world, made a point of telling people never to mutilate trees in any way, as they are a source of life. The text goes on to warn those who contaminate natural resources in any way that they do so at great karmic peril to themselves.

The *Ādipurāṇa*, the epic poem about the first Tīrthaṅkara, Ṛṣabha Dev, underlines the importance of forests. The text emphasizes that forests moderate the climate, check thunderstorms and floods, protect the neighboring areas from cold winds, and enable the constant flow of rivers. They provide shelter for wildlife and fodder for animals, innumerable industrial raw materials, and thousands of excellent medicines. They regulate the underground water levels and provide panoramic beauty. The *Ādipurāṇa* says that forests are like saints, or *munis*, who, overcoming all obstacles, create a better welfare for all. Forests remove fatigue, and every type of life feels better off because of the unique type of ecosystem they create, consisting of trees, plantations, animals, air, and water. The forest is the basis for survival and a symbol of happiness. As in the relationship between bride and bridegroom, it is the duty of all of us to protect and preserve the forest. To live a peaceful life and earn positive *karma*, the *Ādipurāṇa* suggests the planting of a tree. It is said that one who plants a tree remains steadfastly close to God.

Respect for All Life

People often think that religions which endorse forms of world renunciation hold the position that personal salvation supersedes all other concerns and that the search for otherworldly rewards overrides a commitment to this world. Some may think that divine-human relations are considered more important than the relationship between humans and the natural world. This may be true for those who have not experienced a "oneness" with all other souls, a connection of totality. However, *sarvajñyata,* the perfect knowledge of Mahāvīra or of any Tīrthaṅkara, is not only defined as the knowledge of past, present, and future, but also as an extraordinary capacity for experiencing the pain and pleasure of all beings. Tīrthaṅkaras have attained this great state of empathy by the purification of their soul. Once one obtains purification at the highest level, one can realize the infinite knowledge, peace, and beauty, not only of one's own soul, but also experience and have perfect knowledge of other souls as well.

Mahāvīra's campaign for the protection of ecology and nature was not just directed toward the preservation of human life. It was based on his direct concern for the pain the animal and vegetable worlds suffer. He preached his philosophy far and wide and urged his disciples to go and protect animals wherever required.

The *Triṣaṣṭiśalākāpuruṣa*, the famous epic about sixty-three great people in Jain history, recounts that Indrabhuti Gautam, the chief disciple of Tīrthaṅkara Mahāvīra, was sent to stop Agnisharma Brahmin from performing a great sacrificial ritual for which the killing of thousands of animals was planned. Timing was crucial here. When Lord Mahāvīra sent Gautam to prevent Agnisharma's act of violence, he himself was about to attain *nirvāṇa*, the final death, after which Mahāvīra would not have been able to give any more less lessons about perfection to his disciples. Certainly, Gautam would not get the chance to sit at the feet of his master. Yet, Mahāvīra preferred to take a step that was more universal, a feeling more concerned about the life of others—nonhuman others—than to work further toward the spiritual benefit of one single individual. Thanks to the efforts of Indrabhuti Gautam, the slaughter of animals was stopped, and Agnisharma took a vow not to kill animals in the future.

Resource Limitations

Unquestioned and unplanned technological advances are considered to be major causes of the ecological imbalance. Tīrthaṅkara Mahāvīra was also very thoughtful in guiding his disciples in the selection of their trade, business, or profession. He said the people should not go into professions that destroy nature and perpetuate violence. A Jain layperson is required to take twelve vows for spiritual progression. The seventh vow, *bhogopabhoga-vrata,* involves limiting one's use of resources. This has been understood as clearly forbidding involvement in fifteen trades, such as earning a livelihood from the destruction of plants; a livelihood from carts, including the construction and sale of carts to be drawn by animals and then driving them, whether done by oneself or at one's instigation; any trade in animal by-products; trade in lacquer or trade in alcohol and forbidden substances; and trade in destructive articles like poisons, weapons, spades, all of which are potentially dangerous to life.[3] The underlying concept is avoidance of injury to animals, insects, or plant life. Jain teachers were conscious that these professions are potential health hazards because they are major pollutants of the environment. Rejection of such trades also requires that Jains not consume the products of these trades.

Limiting resources and possessions is one of the major vows that the Jain laity observe. Obviously, renouncers are required to take this vow as well. Wants should be reduced, desires curbed, and consumption levels kept within reasonable limits. Accumulation of possessions for personal ends should be minimized. Using any resource beyond actions of necessity, or the misuse of any part of nature, is considered a form of theft, a transgression, and repentance is required.

The Jain way of using resources is beautifully illustrated in the following story, which may be familiar to some readers. Once upon a time, six friends went out together. After a while they were hungry and thirsty. They searched for food for some time and finally found a fruit tree. As they ran to the tree, the first man said, "Let's cut the tree down and get the fruit." The second one said, "Don't cut the whole tree down, cut off a whole branch instead." The third friend said, "Why do we need a big branch?" The fourth friend said, "We do not need to cut the branches, let us just climb up and get the bunches of fruit." The fifth friend said, "Why pick that much fruit and waste it? Just pick the fruit that we need to eat." The sixth friend, quietly, "There is plenty of good fruit on the ground, so let's just eat that first."[4] This shows how destructive one can be if one fails to think through the consequences of one's actions and consider possible alternatives.

In sum, the way Tīrthaṅkara Mahāvīra lived his own life provides one of the most profound examples one can follow in restoring the ecological balance. He used resources sparingly, ate just enough to survive, and had no dwelling of his own. He had no possessions whatsoever, to the extent that he was completely naked. In this way he developed a unique oneness with all facets of the environment. Tīrthaṅkara Mahāvīra's life also shows that progress along a spiritual path does not forbid someone from being concerned about the environment and the world around him, since the two are not mutually exclusive.

Nonviolence and Ecology in Practice

At the root of the Jain path of purification is the concept of *ahiṃsā*, nonviolence. On the one hand, we are taught not to commit violence, and on the other hand, one needs to carry out the positive aspects of the world—that is, to actively promote peace, reverence, justice, and tolerance for all in the world. Tīrthaṅkara Mahāvīra was exemplary in manifesting the law of nonviolence in its entirety. He avoided knowingly committing violence his entire life, and at the same he time maximized his efforts for the protection of all forms of life.

We all know about the environmental and ecological crisis in the developed and developing countries, due to accelerating industrial and technological growth, but there is another form of ecological crisis taking place in the underdeveloped countries. This is the lack of respect for the essentials of life. Natural resources are overused and exhausted just to meet the basic, minimal needs of life. I come from Bihar, the place of Tīrthaṅkara Mahāvīra's birth, teaching, and final death. Most of the population in Bihar lives below the poverty line. The villagers cut down trees for fuel for cooking and sell wood as a way to earn a livelihood. There is a community called Mushara whose food consists mainly of rats. They have nothing else to live on. They dig holes in the fields and fill them with water: the rats in the fields then fall into the pits and drown. The rats are then eaten by the community. The children of that area have no facilities for education and play; they become adept at killing birds and fish. The lakes and ponds have been emptied of fish and other animals in order to feed the hungry in the surrounding area. Sacrificial rituals are still the norm for the local population. This results in the killing of thousands of animals. Needless to say, if such activities continue, the ecological and environmental crisis in such places will become only more devastating.

Veerayatan, a Jain institution, begun in Bihar and run by Jain nuns, is an example of a group that is taking preventative measures to save the environment. Thousands of trees have been planted in the area where Tīrthaṅkara Mahāvīra spent fourteen rainy seasons. People are being given incentives to plant more trees. Drinking water, food, shelter, and employment facilities are being provided for the local population so that their dependence on the remaining natural resources for their livelihood is reduced. Above all, villagers are taught moral and ethical values, including the protection of animals, the phasing out of their use of sacrifice, and the importance of protecting natural resources. As a result, thousands of people in the area have given up their habits of drinking alcohol, killing animals, and eating meat. They have become vegetarians, and they are provided with decent employment supportive of lives at peace with the environment.

This is the prime example of practicing Mahāvīra's compassion for all in our own time. It clearly shows the concern of the religious community and the active involvement of its saintly people in the betterment of our surroundings. A life of renunciation, living in a manner friendly to the environment, strengthens the ecology. Those who follow this way of life engage in actions that are devoted to the protection of their spiritual development, which decreases violence in the world.

The compassionate mission of the people affiliated with Veerayatan is being enacted on a daily basis. Religion, particularly the religion of the nuns, is often seen as passive and unconcerned with the world. In Veerayatan, the reverse is true. The activism at Veerayatan is based on the universal principle of Tīrthaṅkara Mahāvīra that the sun, the air, the water, and nature as a whole give of themselves silently and selflessly all the time. It would be selfish on our part if we take and do not at least return a portion of what we have taken by the time we leave the world. The mission works on this motto, given by Mahāvīra in his last sermon, recorded as the *Uttarādhyayana Sūtra*: "Let friendship be our religion, not only in our thoughts but in our actions as well."

Notes

1. *Ācārāṅga Sūtra* 1.1.5. See *Jaina Sutras*, Part 1, *The Ākārāṅga Sūtra. The Kalpa Sūtra*, trans. Hermann Jacobi (Oxford: Clarendon Press, 1884), 9–11.

2. For the original prayer, see Muni Ratnasenvijay, *Caityavandan Sutra Vivecana* (Falna, Rajasthan: Sarda Prakasan, 1983). For another English translation, see John E. Cort, "The Rite of Veneration of Jina Images," in *Religions of India in Practice*, ed. Donald S. Lopez, Jr. (Princeton: Princeton University Press, 1995), 328.

3. For a full listing of acceptable livelihoods, see Padmanabh S. Jaini's chapter in this volume, and R. Williams, *Jaina Yoga* (Delhi: Motilal Banarsidass, 1983), 117–23.

4. For another version of this story, see Jagmanderlal Jain, *The Outlines of Jainism* (Cambridge: Cambridge University Press, 1916), 47.

Ecology and Spirituality in the Jain Tradition

BHAGCHANDRA JAIN 'BHASKAR'

Introduction

Ecology is the study of animals and plants and how they interact with human beings and their general surroundings or environment. The term "the environment" describes the sum total of physical conditions that influence organisms. It is interrelated to all that lives. Human ecology describes and relates to human behavior toward all forms of life, including plants and animals.[1] Because it recognizes the interdependence of all life-forms, human ecology may be seen as intertwined with the same principles that teach us the lessons of humility, spirituality, and nonviolence.

Jainism, one of the most ancient religions of the world, espouses a nonviolent and humanitarian perspective on all souls. Jainism originated and developed on Indian soil, containing a profoundly progressive social attitude. Even in antiquity, Jain thinkers discussed at length how one could protect the environment and save oneself, society, nations, and all creatures from natural calamities through nonviolence and nonpossession. This chapter attempts to illuminate this subject matter and explain how Jainism has addressed questions of ecology and the environment and created an awareness regarding the interconnectedness and interdependence of all natural systems.

Jainism's Nonviolent Attitude toward Nature

Under the influence of materialistic desires, humans have become involved in an unnatural treatment of their environment, one which overlooks ecological values and opens the way to a myriad of difficulties. Friedrich Engels was right when he wrote, in "The Part Played by Labor in the Transition from Ape to Man": "Let us not, however, flatter ourselves over-much on account of our human victories over nature; for each such victory, nature takes its revenge on us. Each victory, it is true, in the first place brings about the result, but in the second and third places it has quite different, unforeseen effects, which only too often cancel the first." As a result, the disorder of nature has attracted the attention of political, religious, and spiritual leaders who urge us to recognize that disrespect for the environment has come to dominate social values and to consider whether these destructive patterns can be corrected through greater attention to spiritual values. In replying to such questions, human ecology establishes the relationship between humans and nature, between people and the environment in which they live.

The global ecological crisis cannot be solved until a spiritual relationship is established between humanity as a whole and its natural environment. Jains have been staunch protectors of nature since the inception of the Jain faith. A religion of nature, Jainism paves the way to understanding nature's utility and the essential nature of plants, worms, animals, and all sorts of creatures that have their own importance for maintaining ecological balance. Jainism therefore says that the function of souls is to help one another, "*Parasparopagraho jīvānāṃ*."[2] Through an examination of the principles of economics and geography, this principle is connected to the whole of life. It includes humans and other creatures. The plant, animal, and human populations (including their habitats), when seen from the perspective of geography, are merely part of the landscape. For Jains, the landscape itself lives and breathes and merits protection.

Spirituality and Ecology

Spirituality is essentially an individual endeavor; individuals create collectivity on the basis of discipline and practice. Every basic reality of the universe is integral. Jainism reconciled the parts of reality with the whole by means of a relativistic approach. The *Āyārāṅga Sutta* (*Ācārāṅga Sūtra*), an ancient Jain text, therefore says, "one who knows one comes to know all. One who knows all, knows one" (*Je eke jāṇai te savve jāṇai; je savve jāṇai te eke jāṇai*).[3] Kundakunda (first century C.E.) and other Jain thinkers have also followed the same view. This concept emphasizes that everything has innumerable characteristics that can be perceived from numberless instances. Spiritual relationships, from an ecological perspective, can be understood with the help of some of the basic tenets of Jainism: 1) injure no creatures (*savve pāṇā na hantavvā*); 2) do not command any creature; 3) do not own any creature; and 4) do not employ one as a servant (*savve pāṇā na pariggahetavvā*).[4]

Jain ecology is based on spirituality and equality. Each life-form, plant, or animal, has an inherent worth and each must be respected. Within Jainism, a term for ecology might be *sarvodayavada*, or a concern for lifting up all life-forms, as articulated by Samantabhadra (third century C.E.), a prominent Jain philosopher. Ācārya Jinasena explained the same view of social equality by saying that the entire human world is one because of the interconnectedness of different aspects of the human community.[5] Seeing other people as connected with oneself develops a spiritual perspective through which all life takes on sanctity that can and must be protected by observing the principles of ecology. The real task of religion consists in removing bitterness between people, between races, between religions, and between nations. That nature of religion has been discussed in Jain scriptures in various ways and can be summarized:

> Aspire for yourself. Do not aspire for others. This is the fundamental principle of Jainism.
>
> *Jaṃ icchasi appaṇatto, jaṃ ca ṇaṃ icchasi appaṇanto*
> *Taṃ iccha parassa vi ya, ettiyaṃ jiṇasāsaṇaṃ.*[6]

Because one sees others as not different from oneself, an ecological view can arise that sees sanctity in all other persons as well as in plants and animals.

Coordinating Theological and Ecological Perspectives

The theological perspective of each person's religion influences one's treatment of the environment. If, in one's theological teachings, the universe is created by God, and God created the human person in his own image, then God will rule the world through the human person. Humankind is accordingly subordinate to God and functions and rules over nature under God's direction. This creates subordinating ecological attitudes toward nature, including matter and animals. It might also include the notion that nature does not possess a soul or, in the instance of Aristotle, that animals possess inferior souls. In some teachings, the soul is not immortal. In other theologies, body and soul might constitute a single entity; when the body dies, the soul dies.

Jainism does not favor these various views. In the Jain perspective, the soul is eternal and beginningless. It has life, consciousness, knowledge, and perception; it is potent. It performs actions and is affected by their results; it is conditioned by its own body, is incorporeal, and is ordinarily controlled by *karma*. In addition, it is capable of becoming free from the defects of *karma* and achieving salvation. Thus, soul is an independent element which can achieve the highest point of purification and can even become *paramātman*, a great liberated soul. Jainism also holds the view that there is no God who acts as a creator, protector, and destroyer. It is one's deeds alone that bestow the results in our lives, which other traditions ascribe to fate or to the will of God. According to Jainism, the creation of the universe depends on our efforts. For instance, in the manufacture of a pot, clay is the substantive cause, and the potter, the stick, and the wheel are subordinate causes. Each and every entity runs through these two modalities of causation. Accordingly, all entities have their own independent existence: humankind is not the controller or overlord of creation. If one entity overpowers another, problems arise as a direct result. This is precisely what has happened to nature.

Jainism holds that the entire world, including plants, trees, birds, animals, water, and so forth, is possessed of life. It is our prime duty to protect all this. We are to treat others as we want to be treated, and this refers not only to other people but to the entirety of our planet. One is therefore expected to respect the land and its natural beauty. Jainism does so philosophically by accepting the principle of the interdependent existence of nature and animals.[7]

The Concept of Spirit and Its Relationship with *Karma*

According to the Jain faith, to obtain spiritual happiness and perennial peace, it is necessary to believe in the existence of a spirit or soul, which is endowed with such qualities as consciousness, the power of cognition, formlessness, agency, being the enjoyer of the fruits of *karma*s, beginninglessness, infinite in number, and so forth. The soul is of two types. One is worldly (*saṅsārī*) and the other is emancipated from *karma*s (*mukta*). Worldly souls attract an influx of karmic matter and become attached to false notions and false passions. As a result, the soul becomes obscured. This soul is termed *bahirātman*, and it is a soul which is ignorant of the potential for ecological harmony. The second state, called *antarātman* repents of these indulgent tendencies and determines to see the spiritual power within. The third stage is the *paramātman*, the *siddha*, a soul that has freed itself from the bondage of *karma*s and the cycles of death and rebirth.[8]

From an ecological perspective, we should understand the transmigration of souls (*saṅsārīs*), which are of two kinds: mobile souls (*trasa*s) and immobile souls (*sthavara*s). The *sthavara*s are further divided into five kinds—earth, water, fire, air, and plants—each possessing one sense of touch (*sparśa*). The mobile beings are from two-sensed beings onward. The worm, the ant, the bee, the human each have one more sense of taste, smell, sight, and hearing, respectively, than does the preceding order.[9] These living beings possess some essential characteristics, called *prāṇa*s, such as power (*bala*), sense (*indriya*), energy (*vīrya*), longevity (*āyu*), and breath (*ucchavāsa*).

Nonviolence, Religion, and Ecology

Nonviolence and religion are integral and cannot be separated. Religion and spirituality guide human society and show individuals the path to a purposeful life that is respectful of others. Nonviolence depends on experiences that work toward the unity of society and of the cosmos. None would wish injury to oneself. If no sort of self is different from me, how would any one kill any sort of body? The *Ācārāṅga*, *Samayasāra*, and other Jain works clearly say:

> That which you want to kill is yourself. That which you want to satisfy is yourself. That which you want to torture is yourself.

Tumaṃsi nāma sacceva jham' hantavvam'ti mannasi, tumaṃsi nāma jam' ajjāveyavvam'ti manasi, tumaṃsi nāma sacceva jam' uddaveya-vaṃ'ti mannasi.[10]

Thus, nonviolence creates identity between self and self. Therefore, Mahāvīra says, "kill no creature." One has to experience personally the consequences of one's own *karmas* (*Aṇusamveya namappāṇeṇaṃ, jaṃ' ahantavvaṃ'ti ṇābhipatthae*).[11] Through this unitive experience, the existence of souls is established. Mahāvīra goes on to say that one who is afflicted with lust is bereft of knowledge and perception. Truth will always baffle such a person. He indulges himself in action, causing violence to the beings of earth body, water body, fire body, vegetable body, and others. These beings have consciousness (*santi pāṇā puḍho siyā*).[12]

A passage from the first chapter of the *Ācārāṅga Sūtra*[13] clearly describes the cause of violence as the use of weapons. According to the *Ācārāṅga*, passion is the cause of mental perturbance. A perturbed person is easy prey to desire and indulges in violence. Several kinds of weapons are used in violence, such as plowing and digging into the body of the earth, which, as we have seen, harbors many feeling souls. Other weapons are used in a wide array of activities for the sake of survival, praise, honor, reverence, liberation, or prevention of misery. According to Jain philosophy, the earth contains conscious beings who experience pain. One who uses weapons on the body of the earth has neither comprehended nor foresworn actions that cause violence. One should, therefore, not use any weapon causing violence to the being of the earth, cause others to use such weapons, or approve of others using weapons in this way.

The *Niryukti*, or commentary, on the *Ācārāṅga* explains this statement. It says that the process of respiration is present in the earth-bodied beings, but it is not perceptible. Like a person who has fainted, the beings of earth body do not manifest consciousness, because of the constant coma of the deep slumber-producing *karma* known as *styānagriddhi*. Plowing, digging, excretions, and other violations of the earth fall into the category of violence toward the earth-bodied beings.[14] According to the *Mūlācāra*, the earth body includes thirty-six types of soil, sand, gold, silver, stone, diamond, salt, iron, and other products of value to humans.[15]

Likewise, water, according to Jainism, contains beings. This is supported by the tools of contemporary science, which demonstrates the existence of multitudinous microorganisms in water. Jains are supposed to drink filtered water. The *Niryukti* enumerates some weapons that kill the beings found in water, such as drawing water from a well, straining it, washing clothes. These can be seen as heterologous weapons and homologous weapons. There are also weapons that cause violence to fire bodies—soil, sand, water, vegetation. There are beings residing in earth, grass, leaves, wood, dung, and garbage. There are also beings that fly in the air and drop down from the air. All beings shrivel up on coming into contact with fire and experience instantaneous death.[16]

A comparison can be made between plant life and human life. As the human being is born, grows, possesses consciousness, takes nourishment, is not eternal, has metabolism, undergoes transformation, sleeps, experiences longings during pregnancy, and succumbs to disease, so also do plants. Humans do so for the sake of survival, for the sake of praise, and for the sake of honor, as we have already said.[17] Plants, it may be surmised, experience pain when struck or cut with a weapon, just as the consciousness of a person born with sense organs feels pain.

Water-bodied beings are of several types, including dew, hail, snow, and brine,[18] which reside under and above the earth in wells, rivers, pits, tanks, oceans, and lakes.[19] Fire-bodied beings are to be saved by a follower of nonviolence. One should not extinguish the fire by throwing sand or water on it. Likewise, air-bodied beings can be killed by electric fans and even by palm fans.[20]

This is a vivid description of how plants and other beings can and should be saved by a Jain. These are very important principles laid down by Jainism to avoid harm to water, air, fire, and all forms of life and to minimize such evils as sound pollution, and thus to balance the community and ecosystem. To keep silence, to observe carefulness in speaking (*bhāṣāsamiti*), to protect the forest and plants—these are religious rules that apply even to the Jain laity. The bodhi tree, the fig, pepper, mango, plantain, betel nut, bamboo, margosa, palm, tulsi, kadamba, teak, tamarind, acacia, and other trees are very useful for humankind. They should not, therefore, be cut down and thrown away. Jainācāryas have described different types of plants and their classification in several ways. The *Bhagavatī Sūtra*, the *Aṅgavijjā*, the *Gommaṭasāra Jiva Karmkāṇḍa*, and the *Kalyāṇakāraka* are prominent books which deal in detail with the life of plants. Indians worship plants on many different occasions. If plants or trees are cut down, many environmental problems arise. Therefore, Jain tradition forbids the Jain to cut down trees,[21] and there is a vow to refrain from avoidable actions (*anarthadaṇḍavrata*) specifically for the purpose of protecting plant life.

Jain cosmology gives paramount importance to mountains, rivers, trees, and other natural resources. Jain Tīrthaṅkaras and *ācārya*s received their spiritual attainment meditating under trees, on mountain heights, and on the banks of rivers. These natural resources have been enshrined as sacred in Jain art and architecture.

Negligence is the main cause of violence. Even violence in thought, according to Jainism, is a cause of injury. Such texts as the *Daśavaikālika Sūtra* and the *Mūlācāra* say that no sin attaches to one who walks, stands, sits, sleeps, eats, and speaks with vigilance and nonviolence.[22] The *Daśavaikālika* further says, "killing horrifies because all beings wish to live and not be slain."[23] Therefore, nonviolence should be observed strictly, at any cost.

Social Attitudes and Ecology

Ecology sees the individual as interconnected with both nature and the fabric of society. Ecological theory considers the community a supra-organism, a complex social organism. An ecological system represents the highest organizational stage of living matter. Cells form tissues; tissues form organs and limbs, which form individuals. Individuals gather in interrelated communities, forming an ecosystem. The Jain tradition instructs the Jain laity to keep the community very pure and pious. An individual is a unit of the society. Therefore, Jainism stresses individual purification as the foundation for social purification. To this end, Jain laity are supposed to refrain from indulgence in seven types of obnoxious habits (*vyasana*s), which make life disastrous: gambling and racing (*dyūta*), meat-eating, alcoholic drink (*madyapāna*), prostitution, hunting, stealing, and sexual intercourse with anyone other than one's spouse.[24] A Jain should be a strict vegetarian. A Jain layperson should not indulge in professions related to violence, such as dealing in weapons. (A fuller description of acceptable professions is given in Padmanabh S. Jaini's chapter in this volume.) Ideally, one's sources of income should not involve people in antisocial activities. A. L. Basham notes that a truly observant Jain should not be a farmer, for plowing the earth involves injury to animal life. Most crafts involve injury to living beings; for example, as we have seen, the metal on the anvil of the blacksmith suffers excruciating tortures. Thus, the safest profession for the Jain is trade, and from the earliest days of faith, Jainism has recruited its members from the trading communities of India.[25]

Jain laypersons also practice truth (*satya*), not stealing (*acaurya*), refraining from all illicit social activities (*brahmacarya*), and nonpossession (*aparigraha*) to perfection. Avoiding infractions (*aticara*s) by observing the vows keeps the person and social environment pure. Pūjyapāda enumerates five infractions of the minor vows of nonviolence (*ahiṃsāṇuvrata*), such as binding, beating, mutilating the limbs of, overloading, and withholding food and drink from animals. The infractions of the minor vow of truth (*satyāṇuvrata*) are: 1) perverted teaching (*mithyopadeśa*); 2) divulging what is done in secret (*rahovyākhyāna*); 3) forgery or preparing false records in order to cheat people (*kūṭalekhakriyā*); 4) misappropriating or taking for oneself gold and other valuables entrusted to one's care by another (*nyasāpahāra*); and 5) proclaiming others' thoughts (*sākāramantrabheda*). This vow can be strengthened by giving up anger, greed, cowardice or fearfulness, and jesting and by speaking in a manner that will cause no harm.[26]

The infractions of *asteyavrata* (not taking the property of others, whether pledged or dropped) are: 1) prompting another to steal (*stenaprayoga*); 2) receiving stolen goods (*stenahṛtadāna*); 3) underbuying in a disordered state (*viruddharājyātikrama*); 4) using false weights and measures (*hīnādhikamānonmāna*); and 5) deceiving others with artificial or imitation goods (*pratirūpakavyavahāra*).

Celibacy, the fourth vow, is called *brahmacaryavrata*. It seeks to avoid the harm to minute beings caused by sexual intercourse. It also serves to discipline the mind.

Parigrahaparamāṇuvrata is limited acquisition for personal use. This was a formula devised by Tīrthaṅkara Mahāvīra for social reconstruction. It includes the purity of earning sources and an emphasis on austerity in personal life. One is forbidden to indulge the following practices: 1) adulteration of goods; 2) underweighing goods; 3) supplying poorer quality products than promised by a sample; 4) overloading beasts of burden; and 5) harming another's means of livelihood. One should also not deprive native inhabitants of their rights, and therefore Mahāvīra prescribed a vow not to visit remote places for the purpose of acquisition.[27] These vows assist us in eliminating corruption from society and in purifying ourselves in the process.

Nonviolence is based on the principles of equality and equanimity as applied in society. Nonviolence still may allow for a theory of caste, but one based on one's own deeds and not on one's birth. Mahāvīra said that one ought to shun all vanities in knowledge, austerities, caste, and livelihood, as they lead to disrespect for others. One who is free from these vanities and transcends caste altogether achieves the supreme state of casteless deliverance.[28] Mahāvīra further says that caste is like the slough of the human body in that it blinds a human being: "He regains his sight only after discarding the slough. That is why I exhort the saint to shed caste, as a snake does its slough. He should never take pride in his caste. He should never be scornful to anybody on that account. . . ."[29]

The principles of love and vegetarianism are integral parts of ecology. The vegetarian diet addresses the minimum requirement for human life and good health. The will to live at the expense of another's life gives rise to meat consumption, which cannot be acceptable within a truly humane society.[30] The view that a vegetarian diet does not provide sufficient energy or stimulation for active people is simply not correct. It is now an established fact that vegetarian food is more powerful and less pernicious than a diet that includes meat. Tīrthaṅkara Mahāvīra's first motto is that, if violence under compulsion is unavoidable, at the very least voluntarily performed violence is to be avoided. This dictum renders a nonvegetarian diet vulnerable to moral criticism.[31] Therefore, Jainism has laid down many rules and regulations about food, even for the laity.[32]

The Jain tradition is one that considers ecology an indispensable part of both spirituality and material life. Respecting nature is a realistic approach to religion, to life itself: we must realize the paramount importance of ecology and nature for spirituality and the value in protecting nonhuman life-forms for the sake of human welfare. Jainism tries to shape our attitude toward nature by prescribing humane and nonviolent approaches to everyday behavior. Jainism inspired its followers to safeguard what in contemporary discourse would be called the ecological perspective. Jains even today practice these principles and religious traditions prescribed for the protection of nature. Through its philosophy, its ascetic practices, and in its narrative arts and architecture, Jainism and its leaders have made efforts to create a society dedicated to love for all creatures.

Notes

1. James A. Quinn defines human ecology as a specialized field of sociological analysis that investigates 1) impersonal subsocial aspects of communal structure, both spatial and functional, which arise and change as a result of the interaction between humans and the limited supplies of the environment; and 2) the nature and forms of the processes by which this subsocial structure arises and changes. See Quinn's discussion of Hollingshead's "Community Research: Development and Present Condition," in *Studies in Human Ecology*, ed. George A. Theodorson (Evanston: Harper and Row, 1961), 140.

2. *Tattvarthā Sūtra* 5.21. Published in the *Śramaṇa Siddhānta Pāṭhāvalī* (Jaipur: Kunthuvijaya Granthamālā Samiti, 1982), 268. Hereafter cited as *HSSP*.

3. *Pavayaṇasāro, Gāthā* 47, in *HSSP*, 476. See also Niyamasāra of Kundakunda, *Gāthā* 168, in *HSSP*, 513.

4. *Āyārāṅga Sutta*, ed. Yuvācārya Mahāprajña. English translation by Muni Mahendra Kumar (New Delhi: Today and Tomorrow Publishers, 1981), 175. See also *Āyārāṅga* 1.4.23, p. 193.

5. *Manuṣyajātirekaiva jātināmodayodbhavā vṛttibhedāhitadbhedāccaturvidhyamihasnute*. *Jinasena's Ādipurāṇa*, ed. Pannalal Jain, 2d ed. (Delhi: Bharatiya Jnanapitha, 1998), verse 38.45.

6. Samkalaka Jinendra Varni, *Samaṇasutta* (Ravadi: Agama Prakashan, 1996), *Gāthā* 24, p. 7. This *gāthā* emphasizes: "What you desire for yourself, desire for others too. What you do not desire for yourself, do not desire for others." This is the fundamental principle of Jainism.

7. *Vyākhyāprajñapti* (*Bhagavatī Sūtra*), ed. Amarmuni (Beawar: Agama Prakashan Samiti, 1982), 2.2.5.6–7. See also *Prañāpanā* (*Pannavana*), ed. Mishrilal (Beawar: Agama Prakashan Samiti, 1983), 2.157; and *Pañcāstikāya*, in *HSSP*, 527.

8. *Mokkhapāhuḍa*, *Gāthā* 4, in *HSSP*. See also *Kārtikeyānupreksha* (Agas: Rayacandra Granthamala, 1960), 192.

9. Gommaṭasāra Jīvakāṇḍa, *Jīvasamāsādhikāra*, in *HSSP*, 75–77.

10. *Ācārāṅga Sūtra* 1.5.4. See also *Jaina Sutras*, Part 1, *The Ākārāṅga Sūtra. The Kalpa Sūtra*, trans. Herman Jacobi (Oxford: Clarendon Press, 1884), p. 50, for an alternate English translation.

11. See *Ācārāṅga Sūtra* 1.4.1, trans. Jacobi, p. 36.

12. See ibid., 1.1.1–7, pp. 1–14.

13. See ibid., 1.1.2, pp. 3–5. This text title appears in various forms, including the Prakrit form *Āyārāṅga*, and two various transliterations of the Sanskrit form: *Ācārāṅga* and *Ākārāṅga*.

14. *Dhavalā*, ed. Hiralal Jain (Amaravati: Jain Sahityoddharkara Fund, 1939–1958). See also *Gommaṭasāra Karmakāṇḍa*, in *HSSP*, 23.16.

15. *Mūlācāra*, ed Kailashcandra Jain and Pannalal Jain (Delhi: Bharatiya Jnanapeeth, 1984), *Gāthā*s 206–9.

16. Gommaṭasāra Jīvakāṇḍa, *Jīvasamāsādhikār*, in *HSSP*, 75–77.

17. *Imassa ceva jīviyassa partivadaṇa-mānaṇa-puyāṇaye* (*Āyāro* 5.101, 1.2.21). See *Ācārāṅga Sūtra* 1.5.1, trans. Jacobi, pp. 42–43.

18. *Mūlācāra*, *Gāthā* 210. See note 15 above.

19. *Prajñāpanā* 2.151. See note 7 above.

20. *Ācārāṅga Niryukti*, 170.

21. See *Ācārāṅga Sūtra* 2.4.2.11–12, trans. Jacobi, pp. 154–55. See also *Mūlācāra*, *Gāthās* 213–17 (see note 15 above); and *Daśavaikālika Sūtra*, trans. Kastur Chand Lalwani (Delhi: Motilal Banarsidass, 1973), 4.23.

22. *Daśavaikālika Sūtra* 4.9. See also *Bhagavatī Ārādhanā* (Bombay: Anantakiriti Granthamala, 1989), 784–90.

23. *Padmanandi Pañcavimśatikā* 1.16.32; *Lāṭī Samhitā* 2.47–49; *Vasunandi Śrāvakācāra*, 59.

24. *Sāgāradharmāmṛta of Āśādhara* (Delhi: n.p., 1980), 2.2–16; 4.10–11. See also Amṛtacandra, *Puruṣārtha Siddhupāya* (Lucknow: Jagmandar Lal Jaini Memorial Series, 1933), verse 143, p. 59.

25. A. L. Basham, "Jainism," in *The Concise Encyclopedia of Living Faiths*, ed. R. C. Zaehner (Boston: n.p., 1959), 216.

26. Umāsvāti, *Tattvārtha Sūtra; That Which Is*, trans. Nathmal Tatia (San Francisco: HarperCollins, 1994), 7.25–26, pp. 180–81.

27. *Tattvārtha Sūtra* 7.28–30, in ibid., pp. 181–82. See also Samantabhadra, *Ratnakaraṇḍa Śrāvakācāra* (Bombay: Manikachandra Digambara Jaina Granthamālā, 1925), verse 62; and Āśādhara, *Sāgāradharmāmṛta* (Jabalpur: Sarala Jaina Grantha Bhaṇḍāra, 1957), verse 4.64; and *Upāsakadaśāṅga*, with commentary of Abhayadeva, ed. and trans. A. F. R. Hoernle, 2 vols. (Calcutta: Bibliotheca Indica, 1888–1890), verse 418.

28. *Sūyagaḍaṅga* (Ladnun: Jain Visvabahrati, 1982), 1.13, 15.16.

29. Ibid., 1.2.23–24.

30. *Jahā te ṇa piyam dukkhaṃ, jānia emeva savvajīvānam.*
Savvāyaramuvaytto, atto vammeṇa kuṇasu dayām. Samaṇasuttam 150.

31. Muni Nathmal, *Śramaṇa Mahāvīra* (Calcutta: n.p., 1976), 170–71.

32. For more detail, see Bhagchandra Jain, *Jainadharma aur Paryāvaraṇa* (Delhi: New Bharatiya Book Corporation, 2001).

Jain Ecology

SATISH KUMAR

I learned the principles of Jain ecology from my mother. She had never heard of the word "ecology." She would not have seen things in isolation, or by naming or special terminology. But ecology was implicit in the way she lived. For her, the way of Jains and the way of ecology were synonymous.

Until I was six years old, I was educated by my mother. Just as she was not conscious of the term "ecology," she was not conscious of "educating" me. But, looking back I can say that my mother was perhaps the greatest and most important of my teachers.

We have to learn from nature, mother would say. We humans think that we are very clever, but that is our arrogance. Look at the honeybee: she goes from flower to flower taking only small amounts of nectar from each flower; a flower has never complained that a honeybee came and took all of its nectar away.

What do we humans do? When we see something beautiful or useful in nature, we take it, take it until it is depleted and exhausted. If we were to follow the way of the bee, we would learn to take only a little and be content.

When the honeybee has taken the nectar from the flower, what does the bee do with it? She transforms the nectar into sweet, delicious, and nutritious honey, which has great healing qualities. The greatest teacher of transformation is the honeybee.

The worker bee labors diligently and dedicatedly to show us that, not only should we take little from nature, but what we take should be transformed into something greater than what has been taken—into something that is replenishing, nourishing, and nurturing to life.

Mother said: "The honeybee does not know how to create waste! The bee not only teaches the way to transformation, but also the way to pollination. We do not find the fulfillment of our full potential by ourselves; we depend on each other. Man depends on woman. I depend on you. I am grateful to you, my son, that you come into the world through me. You needed my body to be born. I needed you so that I could be a mother. We humans depend on trees and rain and on the fruits of the earth. We need to work to enhance the relationship between us and all life. This is the essence of pollination."

The great teacher and founder of the Jain religion for our time was Mahāvīra, and he taught us to be like the honeybee. He went from door to door and from household to household, begging for his food, always taking just a little. If someone offered him two pieces of bread, he would only take one. No householders could ever say that the begging monk had taken all their food, that they had been left hungry, that they been forced to cook again. Mahāvīra would go once a day to a dozen different doors to get enough for a single meal.

It was as though he had learned the great virtue of restraint from the honeybee. Mahāvīra was like the bee, a great pollinator of wisdom. He went alone, always walking barefoot, carrying no possessions, not even a begging bowl, and he brought stories and dreams, and myths to peasants and princes, to the poor and the wealthy. Walking the earth was a most sacred act—a form of continuous pilgrimage. There have been twenty-four Jain masters, or Tīrthaṅkaras, in the history of the Jain religion, and they all walked up and down the land of India, covering thousands of miles. The Tīrthaṅkaras, communing with and meditating upon nature, roamed through the wilderness with a deep sense of wonder about the mystery of life. Countless mystical experiences in the natural world brought them enlightenment.

My mother used to say to me, "How do you think the Tīrthaṅkaras became enlightened?" And she herself would answer, "They lived in the wilderness, sat under trees, and communed with nature. How is it that we do not have Tīrthaṅkaras, enlightened beings, these days? Because we have cut ourselves off from nature and live in towns and cities. Even the monks no longer go to the mountains and sit under the trees." The Tīrthaṅkaras spent long periods meditating in caves in the mountains, and all of them went to the mountains to die.

These mountain sites are still sacred to the Jain, but these days Jains go to the mountains only because the Tīrthaṅkaras died there. This is a pity. The mountains are themselves sacred. Of course, those mountains which hold the sacred souls of the Tīrthaṅkaras are dear to us Jains, but we should not forget the fundamental truth that all mountains are sacred and are sources of divine inspiration.

Mother herself undertook pilgrimages to the six sacred mountains: Ābū, Palitana, Śatruñjaya, Girnār, Rajgir, and Samet Śikhara. She also went to the holy mountains of the Himalayas. There is no doubt in my mind that these pilgrimages had a deep and profound effect on her.

When Mahāvīra walked, he walked barefoot, treading lightly on the earth. He kept his eyes on the ground to avoid stepping on any living creature. If, by mistake, he stepped on any form of life, he would cause less harm because he walked barefoot. Not injuring any life was his greatest concern and passion. Such was his reverence for life that he taught his disciples to refrain from eating any kind of meat. His concern for life went even further. He asked his followers not to eat root vegetables because, in order to eat a root, one has to dig up the whole plant: thus, grains, beans, peas, pulses, and fruit are suitable foods, whereas onions, garlic, potatoes, carrots, and other root vegetables are not. The principle of limiting one's consumption is the bedrock of Jain ecology.

Mahāvīra instructed his followers that when they prepared fruit and vegetables containing multiple seeds, such as melon, they should remove all the seeds carefully for resowing and regeneration. That Jains should always save seed was propagated by Ādināth, the first of the twenty-four Jain masters. Ādināth taught that farmers should always sow more than they would need so that there would be enough for the birds, the mice, the ants. When peas, pulses, and grains are harvested, farmers should keep enough seed for sowing the following year and should also leave seed on the ground for wildlife. In times of scarcity, wild creatures should be fed; for instance, my mother would put grain on the top of an anthill.

Such restraints did not mean that food was a dull affair in our home. In fact, each time mother prepared a meal, it was a minor feast. The aroma of spices, the color of foods, the manner of presentation, the ritual of eating, and the care given to each member of the family made every day an occasion of celebration. Mother believed that if you had bad food in your belly, you could not have good thoughts in your head. I remember the delights and pleasures of home cooking: the fewer the ingredients, the greater the challenge to the imagination and the art of cooking.

Mother followed the example of Mahāvīra and always walked barefoot. When we touch the sacred earth and feel the coolness of the soil, we know that we have come from the earth and will return to the earth. Walking barefoot brings you in contact with the soil. There is no better way to health and happiness than a daily practice of walking, in solitude, barefoot upon the earth. The minerals of the earth enter the body through the soles of our feet and heal our souls. Fear and frustration, anger and anguish, tears and tribulation quickly disappear, blowing away with the wind when we walk among the trees and listen to the hum of bees.

Mahāvīra forbade riding on animals. He saw no justification for humans to exploit animals, since human beings have perfectly good legs to use to move around. "Would you like to be ridden by monkeys, with a rein passing through your nose?" So, mother never rode on horse or a camel and would not ride in a cart pulled by an animal if she could avoid it.

Mahāvīra taught his disciples that the earth has soul, water has soul, fire has soul, air has soul, and of course all plants and animals have souls. Just as humans love life and do not wish to be harmed, nonhuman beings, including insects and mosquitoes, worms and spiders, butterflies and bees, yearn to live. Therefore, compassionate human beings should make a constant effort to reduce the damage they inflict on other forms of life.

Mahāvīra's goal was to take as little as possible from the natural world and to live in harmony with nature. His compassion was so apparent that when he preached his sermons, which would always take place in a grove of trees or in a forest, not only humans would come to hear him speak; so too would lions and tigers, eagles and doves, peacocks and bullocks, all sitting at his feet.

Jain ecology insists that we must learn to respect the water body, the fire body, the earth body, and the air body. These natural bodies are not distinct from human bodies; human bodies contain earth, air, fire, and water. These elements sacrifice themselves to sustain the human body. Therefore, it is only right that human beings should take responsibility for sustaining the natural order and preserving the integrity of the elements. This is the principle of *ahiṃsā*, nonviolence. It is the most fundamental principle of Jain ecology. But *ahiṃsā*, or nonviolence, was not just a dogma for my mother: it was a way of life.

One of my mother's favorite occupations was to decorate shawls and skirts, blouses and waistcoats, with mirrors embroidered into the cloth. She was well-known for her skill. Friends and neighbors would bring her nice pieces of mirror for recycling. These she would sort according to size and keep in baskets lined with soft cloth. In other baskets she would keep embroidery silk of all colors. With a needle, plenty of time, and a vast reservoir of imagination, she produced intricate patterns. She saw her embroidery as a process of transformation not dissimilar to the work of the bee making honey. A needle is the smallest tool ever made by human beings, requiring the least amount of metal; and yet, with it mother could produce garments of extraordinary beauty.

I remember my mother giving a shawl once to my sister as a present. Of course, my sister was delighted, but she said, "Mother, it's too beautiful to wear. I might spoil it. I would like to put it on the wall so people can see it and admire it." But mother said, "No, that will not please me. I have made it for you to wear. If it's spoiled or wears out, you should be able to make another one for yourself or I might make another one for you. When people start to put beautiful things on the wall, they start to put ugly things on their bodies."

She talked about three qualities of Jain ecology that should be remembered when we make things for daily use. First, whatever we make should be Beautiful; second, it should be Useful; and third, it should be Durable. I call these the BUD principles. If any one of these three qualities is missing in our work, then we have missed the mark.

It took my mother nearly six months to complete that shawl, but she never worried about the length of time it was taking. In fact, she hardly measured time. My sister said to her, "Mother, nowadays you can buy machines which will sew and embroider." My mother replied, "Why do you want me to have a machine?" "Mother, it will save you time," said my sister. "Is there a shortage of time? When God made time, he made plenty of it. Time is eternal and it is always coming—time is not running out. When you are making something, you must not think of time. What is the hurry? Whatever we make should be made well and not quickly. And if we can't do things in this life, there is always the next life!" "But mother, if you save time you can do more things, make more shawls." Mother said, "But you are trying to save time, which is infinite, and you are using up things which are finite. What kind of logic is this?" "All right then, if not time, it will save you labor, Mother." "Yes, but like time, human labor is also not in short supply. Time and labor are like air and water; they are abundant and free."

Another of Mother's passions was making quilted jackets and bed covers from saris, bed spreads, and other pieces of material that we no longer used. Carefully combining colors, she would transform what was worn and worthless into attractive items of dress. Mother took pleasure in ordinary things and saw in them the potential to be extraordinary. Life itself is a tapestry made up of millions of small acts, each one making an essential contribution to the realization of the meaning of life. She did not believe in acts of heroism. She believed in small actions performed with great love and imagination. She gave to all material things a high value. She saw matter as the vehicle of the spirit and therefore handled all matter with reverence. Work was never a burden for mother; she saw work as a means of self-realization. She believed that the whole of life should be lived as a spiritual practice, as a meditation. Self-realization, or *nirvāṇa*, is not something for tomorrow, not somewhere in the far distance; it is here and now, in all of our actions guided by reverence for matter, reverence for work, reverence for life—in short, *ahiṃsā*. Jain ecology encourages us to live in the world with a sense of the sacred.

As a traditional Jain family we did not live as isolated individuals, we were not taught to stand on our own feet and fend for ourselves. Ours was a relationship of mutuality, mutual sharing, mutual caring, no privacy, no private possessions, no private wealth. Everything belonged to the family. My three brothers worked together in the family business and shared the family home. Nobody knew or calculated who earned what. It was all shared. A family security system operated. The extended family, which included my mother, her sons, their wives and children, as well as my father's brother and his family, lived in a house built around a courtyard. The inner courtyard was connected to an outer courtyard where cows were milked, herbs were grown, rain water was collected, guests were accommodated, and corn was dried, threshed, and winnowed. The whole arrangement was simple and sustainable. Everything was in good order and seemed amply sufficient for our needs. The family business, which was based on the trade of grain and jute, brought in enough cash income to meet the needs of the family. We were not rich, we were not poor, and we never thought of these categories. A sense of satisfaction and "enoughness" prevailed.

An important expression of Jain ecology was the collection of monsoon rainwater in my family. The covered water tank, which was sufficient to hold monsoon rain running off the roof of our house, was as integral a part of our home as the grain stores or the kitchen. This water tank held all our drinking water for the entire year and never failed us.

Most of the households in the town where I grew up had such water tanks. The harvesting of the monsoon rains was one of the major activities of the community. There were small ponds, large ponds, lakes, and step wells everywhere. There were wells ancient and modern, and new wells were always being dug. It seemed to me that the collection, conservation, and preservation of monsoon water and the system to hold it was the most important communal activity. Our share of water was one bucket per person per day. For Jains, wasting water was a serious infraction. Very early on, my mother taught me a water *sūtra:*

> Waste no water
> Don't ever spill it
> Water is precious
> Water is sacred
> The way you use water is the measure of you
> Water is witness
> Water is the judge
> Your wisdom rests on your careful use of water.

Mother often used to say that the monsoon was a great friend of the people and the earth. The monsoon came once a year and brought the gift of water. Our task was to receive the gift with gratitude, thank the rain god, Indra, and make use of water with care and reverence. Our task was to live in harmony with the monsoon and celebrate it. The god Sūrya, the sun, and the god Indra, the rain, are twin brothers and all life depends on them. This was an example of *ahiṃsā* in practice.

Broadly speaking, *ahiṃsā* means avoidance of harmful thought, harmful speech, and harmful action. It involves freeing oneself of any ill will and refusing to entertain any negative thought. It means avoiding contact with scenes of cruelty and refraining from activities that cause pain and disharmony. *Ahiṃsā* enhances goodwill, positive thought, loving action, restraint. Think less, speak less, and do less. Be more, meditate more, practice silence, and serve others. *Ahiṃsā* is much more than live and let live; it is "live and love." *Ahiṃsā* is the first of five fundamental principles of Jain ecology.

The second is *satya*, which means understanding and realizing the true nature of existence and true nature of oneself. It means accepting reality as it is and being truthful to it, seeing things as they are without judging them as good or bad. It means "do not lie" in its deepest sense: do not have illusions about yourself. Face the truth without fear. Things are as they are. A person of truth goes beyond mental constructs and realizes existence as it is. Living in truth means that we avoid manipulating people or nature because there is no one single truth that any mind can grasp or tongue can express. Being truthful involves being humble and open to new discoveries, and yet accepting that there is no final or ultimate discovery. Truth is what is: we accept what is as it is, speak of it as it is, and live it as it is. Any individual or group claiming to know the whole truth is by definition engaged in falsehood. Ultimately, existence is a great mystery.

The third principle is *asteya*, which means refraining from acquiring goods or services beyond our essential needs. It is difficult to know what our essential needs are, so Jain ecology requires that we assess, examine, and question, day by day, what is our need and what is our greed. The distinction between need and greed can be blurred and therefore the examination of need should be carried out with honesty. The principle of *asteya* includes "Do not steal." The Jain understanding of this goes further than any legal definition. If you take more from nature than meets your essential need, you are stealing from nature. For example, clearing an entire forest would be seen as a violation of nature's rights and as theft. Similarly, taking from society in the form of housing, food, and clothing in excess of one's essential requirements means depriving other people and is therefore theft. If we are using up finite resources at a greater speed than they can be replenished, then we are stealing from future generations.

Asteya does not stop with goods and tangibles. If anyone claims to have ownership of knowledge and turns it into a commodity and proclaims intellectual property rights, this is the breaking of *asteya*. Living by speculation or usury, on interest or investments, is theft. It is understood, however, that the whole of society is not committed to the Jain principle of *asteya* and, therefore, even those who adhere to the Jain tradition cannot practice *asteya* in its entirety. *Asteya* is an ideal to which Jains aspire.

The fourth principle is *bramacarya*. This principle has been closely associated with the conduct of sexual behavior: *bramacarya* is love without lust. For monks, *bramacarya* means total abstinence, and for laypeople it means fidelity in marriage. Any thoughts, speech, or acts that demean, debase, or abuse the body are against the principle of *bramacarya*. The body is the temple of pure being, and therefore no activity should be undertaken which would defile the temple. *Bramacarya* not only recognizes the dignity of the human body, but also the body of nature. Brahmā, the pure being, is the essential principle of creation, and nature is the body of Brahmā. For Jains, ecology means living in a relationship of fidelity with nature.

The fifth principle is *aparigraha*, which means nonaccumulation and nonpossessiveness. We do not own anything: everything belongs to itself. There is a relationship between me and the house where I live, between me and the land I farm, but I do not *own* the house or the land. I take enough for my needs and leave the rest to itself. If no one hoards, owns, possesses, or accumulates anything, there will never be any shortage of anything. Nature is abundant. "There is enough for everybody's need but not enough for anybody's greed." *Aparigraha* means sharing. *Aparigraha* means to live simply, without ostentation and without a display of wealth. Dress, food, and furnishings should be simple, elegant, but austere. "Simple in means, rich in ends" is the Jain ecology. When we spend too much time in the care of possessions there is no time for the care of the soul.

Aparigraha means do not acquire what is not necessary. Do not shop beyond your daily need. Recognize that whatever you acquire will bind you tightly. Free yourself from nonessential acquisitions and from materialism.

This means that the acquisition of every pot, chair, or table is a religious decision. Consumerism cannot exist comfortably with *aparigraha*. Practitioners of Jain ecology limit the type and quantity of food to be consumed and the number of clothes to be worn each day.

In Jain ecology each act involves limiting consumption. For a Jain, to shop or not to shop is a religious question. A Jain moves from having to being and from quantity to quality. *Aparigraha* is a form of "holy poverty" or voluntary simplicity. We could even call it "down-sizing," a process of consuming less for the benefit of the self and others.

Tradition and Modernity: Can Jainism Meet the Environmental Challenge?

From Liberation to Ecology: Ethical Discourses among Orthodox and Diaspora Jains

ANNE VALLELY

Introduction

The geographical and cultural distance from India has led to changes in the beliefs and practices of what constitutes Jainism in North America. Traditional orthodox Jain ethics are renunciatory and individualistic, and their central ethic of *ahiṃsā* reflects this ascetic orientation. However, within a growing segment of the diaspora community, Jain ethics no longer reflect the ascetic ideal. Rather than through the idiom of self-realization or the purification of the soul, ethics are being expressed through a discourse of environmentalism and animal rights. Many among the immigrant community enthusiastically and effortlessly embrace "green" concerns, arguing that they resonate fully with the teachings of the Jain tradition. I argue here, however, that the compatibility between Jainism and environmentalism is largely a new, diaspora development, and actually reflects a shift in ethical orientation away from a traditional orthodox liberation-centric ethos to a sociocentric or "ecological" one. This shift away from the ascetic ideal must be viewed as part of the larger theme of immigration and adaptation to a new cultural setting. It, along with the promotion of interfaith activities and nonsectarianism, is an expression of this new social reality. My focus here is on Jain ethics as they have been traditionally understood and constituted in India, and at how they may be changing as they take root in North American soil.[1]

Although the Jain emphasis on nonviolence and compassion would suggest an obvious accord between Jainism and environmentalism, the move toward equating the two represents a deviation from traditional philosophy. Paul Dundas, in this volume, cautions against appealing to decontextualized textual passages in order to establish an essentialized "green" Jainism. He argues that, despite being subject to a variety of interpretations, the ascetic ideal has always been at the center of Jainism. Efforts, therefore, to adjust this philosophy of "world renunciation" to a philosophy that espouses love of, and attachment to, the world are questionable. The shared normative prescriptions of traditional Jainism and environmentalism rest on very different metaphysical views. Yet, this is where modern diaspora Jainism makes a radical departure; its efforts to merge environmentalism with traditional teachings do not rest upon an artificial juxtaposition of opposing philosophies. Instead, they are the outcome of a fundamental reinterpretation of Jainism's ethical orientation. The shift in the understanding of what constitutes "reality" and "human nature"—away from a connection to an ascetic soteriology—is the concern of this chapter.

In spite of their divergent views, however, Jainism and environmentalism do share much in common, and I believe that Jainism has an important role to play in fostering a new social ethic. Indeed, Jain ethics (both in the diaspora and traditionally) are deeply concerned about relationships with the nonhuman environment, reflecting an ontology whereby moral value is constituted, above all else, through interactions with "nature." Its ethical commitment to the avoidance of harm to all life-forms, however this is reasoned, offers an important restorative to views which treat nature as a mere backdrop for human activity, as are common within the Judaeo-Christian tradition.

Background: The Jain Diaspora in North America

In India, the dominant culture of Hinduism has, for the most part, accommodated rather than challenged the Jain worldview. Although

Jainism has its own history, distinct convictions, and traditions, it shares with Hinduism certain core beliefs, such as reincarnation and *karma*, as well as the desirability of asceticism and world renunciation. The situation is, however, profoundly different in North America where the dominant culture does not embrace these core beliefs. In North America, Jainism is being fostered in an environment where the standard markers of identity are either absent or altered. For instance—and importantly—there are no Jain ascetics outside of India.[2] Ascetics act as powerful symbols of the tradition and are typically its central preceptors.[3] Their presence and interaction with the laity assures sect and *gaccha* (ascetic lineage) loyalty, as well as the continuation of traditional, orthodox practices (namely, ritual, prayer, fasting, correct practice, scriptural knowledge). The absence in North America of ascetic leadership appears to allow for the existence of a wider variety of religious belief as well as the development of a nonsectarian Jain socioreligious identity.

The absence of ascetic leadership is likely an important factor in a decline of sectarian affiliation, particularly for second-generation Jains. A number of the community's first-generation immigrants believe that Jainism as practised in Canada is less "authentic" than that practised in India, and readily attribute the perceived "degeneration" of Jainism outside of India to the absence of ascetics. (This group was eager for me to conduct my doctoral research in India where I would encounter "real" Jainism.) Others (especially, but by no means exclusively, second-generation Jains) consider orthodox Jainism to be overly tied to caste and sect and essentially exclusive. They reject these "social" dimensions of the tradition and espouse a universalistic, modern interpretation. They emphasize the values of vegetarianism, animal rights, environmentalism, meditation, and nonsectarianism and actively promote interfaith activities. The majority of the federation of Jain associations in North America and the United Kingdom would fall under this latter category. Thus, for instance, the *Jain Study Circular* (published in New York), the *Jain Digest* (Ohio), *Jiv Daya* (California), and the new international journal *Jain Spirit* (United Kingdom) all champion a modern, scientific vision of Jainism.

The actual number of Jains living outside of India is uncertain.[4] It has been variously estimated from a low figure of seventy to eighty thousand[5] to as high as one million.[6] Despite an insistence upon the exclusivity of their religion, Jains have often displayed a degree of fluidity in religious identification (for example, until recently—and only after considerable campaigning by Jain leadership—it was common for Jains to record themselves as "Jain-Hindu" on Indian census enumerations). That Jains have, in certain contexts, defined themselves as a subsect of Hinduism, or have emphasized caste over Jain identity, demonstrates the complex nature of religious identity and has made the tradition difficult to pigeonhole.[7]

Today, the Jain diaspora is comprised of members from both the Śvetāmbara and Digambara sects; from a wide number of *gaccha*s and castes; and also from all regions of India. At the turn of the twentieth century, however, the vast majority of immigrants, travelling mainly to East Africa, were Gujaratis and Śvetāmbara image-worshipers. Due to political instability in East Africa, Jains began emigrating to Britain and North America in the late 1960s. Marcus Banks estimates that as much as 80 percent of the Jain population of Britain today is Gujarati.[8] In Canada, approximately 50 percent of the Jain population arrived via East Africa; the rest arrived directly from India. Dundas writes that among the first wave of Jain immigrants to East Africa, caste connections were the basis of support networks and played a more important role in the successes of the migrants than did their identity as Jains.[9] However, since the emergence of communities in Western Europe and North America, there has been a strong movement to establish an international Jain community and an inclusive religious identity across caste and sectarian lines.

Jains, like the larger Indian immigrant group of which they are a part, constitute a new community in North America. While Jains have been immigrating here since the late nineteenth century, their numbers until recently were small. No self-perpetuating community existed until the 1970s, when the American and Canadian immigration laws (formerly restrictive to non-Europeans) opened up to Asians. In Canada, two-thirds of all Jains are settled in Ontario, mirroring immigration patterns in general.

In the early 1980s, the Jain Society of Toronto was founded and the first Jain center was established in Toronto. It replaced a home temple (a residence) that had served the community's needs since 1974. The center houses a temple that is used jointly by members of the Śvetāmbara Deravāsī (idol-worshiping), Śvetāmbara Sthānakavāsī (non-idol worshiping), and Digambara sects. In addition, it publishes a newsletter and hosts a wide range of community activities and cultural events.

Although distance from India has led to changes in what constitutes Jainism among the diaspora community, it has not diminished its significance as a potent source of identity. Rather, it may be argued that it has *increased* its importance. The extent to which Jains in India constitute a discrete social entity, transcending sectarian, caste, linguistic, and regional differences, is a matter of debate.[10] In North America, by contrast, there has been a major effort to create such a self-consciously distinct religious identity. Of course, the old fault lines of language and sect remain sensitive—especially among first generation immigrants. But the goal to move toward a nonsectarian, unified tradition that will be an outspoken champion of nonviolence, interfaith dialogue, and environmentalism is a genuine and widely accepted one.

I turn now to a brief discussion of Jainism's traditional liberation-centric ethic in order to contrast it with the emerging socio- and eco-centric ethic of the diaspora community.

The Moral Self in Orthodox Jainism

Through a brief description of a typical "day-in-the-life" of an ascetic (based on my ethnographic experiences of living with a Jain ascetic order), I hope to elucidate the traditional Jain understanding of "nature" and the ideal role of human beings in nature.

> It is three in the morning, and deep within the walls of the monastery, the *sādhvīs* (nuns) begin to wake. The unruffled thin sheets covering their bodies indicate tranquil sleep, an auspicious sign. Completely motionless sleep is the objective, because all movement is potentially dangerous to other living beings. Unkempt sheets would indicate reckless gestures and, thereby, sin. Spontaneous, unmonitored activity leads to the death of countless lives that surround us at all times. The *sādhvīs* rise slowly from lying on their backs and settle into a lotus posture on the cardboard mat upon which they had slept. The *muhpattī* (cloth mouthshield) remains on. It is worn during the night for the same reasons it is worn during the day—to prevent the harm to subtle living beings in the air—and is removed only when eating. Sitting tranquilly, the *sādhvīs* begin their prayers and meditation, awaiting the rise of the sun. In the darkness, where the presence of tiny living beings are concealed, movement is folly.

After more than two hours the first hint of light sneaks in through a window. The countdown is on: "thirty minutes to go . . . twenty . . . ten . . . five . . . ," and finally there is enough light to make out the fine lines in the palm of one's hand, and activity can begin. By insisting on sufficient light to read one's palm lines, the ascetics ensure their ability to be mindful of the living beings in their proximity. They begin the practice of *pratilekhana*—(the meticulous inspection of one's clothes, and all items, for tiny life-forms). Finally, with the sun high in the sky, there is a quick retreat from the building and we are off for the collection of alms. Consuming food is strictly prohibited when the sun goes down, so the items collected this morning will be the *sādhvīs*' first bit of food and drink since before sunset the previous day.

As we move swiftly along the village paths in search of alms, we are mindful of each step. We walk only on sand and cement, for walking on grass would mean killing it; brushing against a bush would mean harming it. Jain ascetics cannot prepare their own food. Plants, water, fire, electricity—all the things necessary for cooking—are considered alive. By the time the ascetic consumes the food, it must be devoid of all life. By ingesting food and water that are no longer alive, the ascetic accrues no *karma*. Through the generosity of the pious householders, the ascetics remain karmically unaffected by violence inherent in the preparation of food. The householder, who has not renounced the world and is living "in" society, accepts that a certain amount of violence is necessary in order to survive and is more than happy to provide ascetics with alms. By doing so, she earns good *karma*.

We approach the home of a pious layperson. Her door is open and she beckons us in. We enter, and the *sādhvī* examines the food before her. Importantly, she asks the woman for whom she has prepared the large quantity of food. The woman explains that it is made for her family—and certainly not with the expectation of giving it as alms to the ascetics. (Lay Jains, the householders, are not allowed to prepare food explicitly for ascetics and should not know in advance that the ascetic will be coming to beg. Instead, the two are to meet by chance. To prepare food explicitly for the ascetics would involve them in violence.)

And how was it prepared? asks the *sādhvī*. The woman of the house explains in detail that all the plants and fruits were boiled; that the water used in making porridge was first boiled. The nuns stand away from the small fridge in the center of the room—for to brush against it would cause harm to fire-bodied beings in electricity. The *sādhvī* asks the woman if she washed her hands with "raw" water. No, the woman is emphatic, she only allows boiled (i.e., "dead") water to touch her. Satisfied, the *sādhvīs* yield their alms bowls and collect a small quantity of food. Then we are off.

After visiting five or six homes, the *sādhvīs* have collected enough food, and we head back to the monastery grounds. The food is distributed among their small group and consumed in its entirety. If any is left over (which doesn't happen often), it must be buried. Discarded food would become the source of an orgy of violence—insects would swarm in it, dogs would eat the insects and would fight among themselves for it—and the ascetics would be implicated in the violence. After their meals, the ascetics depart again (in pairs) for "excretory purposes." Because water is alive, ascetics cannot use flush toilets, so they venture away from their dwellings to find a patch of land that is devoid of vegetation. Again, their refuse must be buried.

After sunset ascetics are prohibited from going outdoors because the night air is filled with dew, which, like rain, is alive. If it is absolutely necessary to go outside, they must cover their heads with a cloth so that the falling dew or rain (water-bodied beings) will hit the fabric, and not die as a result of impacting directly against their bodies. Before sunset, they recite a prayer of *pratikramaṇa* in which they repent for the sin of violence that may have occurred during the day; and they perform *pratilekhna* again before changing into their night clothes. No food or medicines are consumed after sunset and, since ascetics cannot use electricity, they remain in the dark until a householder turns on a light. By nine o'clock most have carefully lowered themselves onto cardboard mats (thoroughly examined for insects) and fallen fast asleep.

The entire logic of the ascetics' daily routine is dictated by the ethic of *ahiṃsā*, or nonharm. It is in interactions with the nonhuman world that the ascetics are most highly attentive, observant, and mindful—that is, when they are most quintessentially ascetic. Interactions with "nature"—with the air, water, soil, and vegetation –define both lay and ascetic Jains by determining the boundaries of their ethical being. The Jain scholar James Laidlaw insightfully describes *ahiṃsā* as an "ethic of quarantine." He argues that Jains' elaborate practices of nonviolence are not so much about minimizing death or saving life as about keeping life "at bay" and essentially amounting to an attempt at the "avoidance of life."[11]

Since all aspects of physical reality are imbued with life, harm done to any one of these lives results in karmic inflow. Violence against other human beings is, according to Jains, obvious wickedness. Such a gross violation would immediately condemn the culprit to countless future lives of suffering, and in an embodiment other than that of the human form. Since violence against human beings is universally condemned and relatively rare (in the sense that it is not something everyone engages in), the true cause of karmic bondage responsible for the cycle of death and rebirth (*saṃsāra*) must lie elsewhere. For Jains, it is in the violence inherent in daily social life that is the cause of our bondage; simply existing leads to the unavoidable death of innumerable beings each day. In the most ancient of Jain texts, the *Ācārāṅga Sūtra* of the fourth century B.C.E., it is written:

> Action, whether done, caused or condoned by oneself, brings about rebirth, and the world is in a state of suffering caused by actions of ignorant people who do not know that they are surrounded by life-forms which exist in earth, water, air and fire.[12]

Rather than being peripheral, interactions with the nonhuman world are at the very center of the creation of the Jain moral self. Jainism treats the whole of existence as part of its moral community. Moral worth is established, above all else, through interactions with the nonhuman world. Jains do not believe in a creator of the universe; they have no belief in a transcendent god, and therefore salvation is not sought through a relationship with the divine. Instead, nature becomes a moral theater, within which one's ethical being is established, cultivated, and judged.[13] This is a central feature of Jain ethical life. And, in spite of other profound differences, it is as equally central for Jains in the diaspora as it is for Jains in India. I will return to this point in the discussion of the Jain immigrant community, but first it is necessary to provide an understanding of traditional, orthodox Jain ontology and its understanding of the ideal role of human beings in nature.

Traditional Jain Ontology

Jainism bases its teachings on a fundamental division of all existing things into two classes: *jīva* (that which has a soul), and its negation, *ajīva* (that which is devoid of soul). *Ajīva* represents physical matter, also called *karma*—as Jains understand *karma* to be a physical, material substance that sticks to the soul. The soul has no form but, during its worldly career, it is vested with a body and becomes subject to an inflow of karmic "dust" (*āsrava*). These are subtle material particles that are drawn to a soul because of its worldly activities. All worldly souls—be they in the form of a worm, a blade of grass, a dewdrop, a clod of earth, or a human being—are in karmic bondage. The soul's association with karmic matter prevents it from realizing its true and omniscient nature, and it will go through a continuous cycle of death and rebirth until it attains *mokṣa*, or spiritual liberation.

Ahiṃsā is the central practice in the quest for liberation because it defines—negatively—a state of purity and detachment within a violent, passionate world. It is an ethic of noninterference and a method of disconnecting or separating oneself from the violence of everyday life. And, in so doing it establishes *difference. Ahiṃsā* makes the human incarnation unique among all living beings by making it moral. We see that in Jainism, the moral self is created through a retreat from the rest of nature. A central focus of all cultural life, cross-culturally, is devoted to establishing what it is to be properly human; the attempt to establish human uniqueness in contradistinction to the "otherness" of the environment is fundamental to human self-definition. The philosopher Charles Taylor argues that a moral reaction is an affirmation of a given "ontology of the human"—in other words, our morals reveal our notions of what it is to be truly human and why humans are worthy of respect. In Jainism, human moral is established through restraint, stemming from a recognition that we share this world with a multitude of living beings, all of whom are now in bondage but are on the same path as we are to eventual liberation.

In the Judaeo-Christian and secular Western philosophical traditions, human dignity resides in that which *distinguishes* us from the nonhuman environment, in particular, that which *distinguishes* us from animals. Animals have always served as a contrast to illuminate human nature, and human worth is located in those areas that we believe we have a monopoly on—for example, a soul, rationality, language, morality. In Jainism, human dignity and moral consideration are not rooted in a nature-culture distinction. Value is not dependent on that which nonhumans lack. Although Jains treat the human embodiment as a privileged and exalted one, they do not believe that humans possess anything *uniquely* or *exclusively* that should entitle them to their superior status. There exist five categories of living beings in Jain cosmology, each being having either one, two, three, four, or five senses.[14] The greater the number of senses, the greater the self-awareness and, therefore, the greater the ability to understand worldly existence as a state of bondage in need of escape. But, an increase in the number of senses does not mean greater moral worth. It is the possession of a soul, and not the stage of development nor the number of senses a being possesses, that entitles one to a life of dignity and respect. Therefore, moral consideration does not hinge upon that which the human incarnation possesses alone, but on that which it shares with all other beings.

Consciousness is the inalienable characteristic of every *jīva*, however undeveloped it may be. It is present even in the *nigoda*s (the least developed life-form) and, through its progressive development, the *nigoda* too may culminate in the supreme state of the soul, namely omniscience. Jains believe that the soul passes through an infinite number of embodiments as it progresses from the lowest to the highest state of spiritual development.[15] These states have been classified into fourteen stages called *guṇasthāna*s, each of which is necessary to pass through to attain liberation.

In the Jain universe, plants and animals are not believed to have autonomous existences, but rather, form part of the same tragic drama of bondage and liberation that humans do, albeit with fixed roles and characters. The whole of existence is assumed to be ordered in a hierarchical scale, moving up to the Tīrthaṅkaras (the revered teachers) and down from them, in what is regarded as diminishing degrees of perfection to ascetics, *śrāvaka*s (laypeople) animals, plants, and single-sensed beings. Plants and animals are considered to be moral symbols of inherent pedagogic value and are judged according to the same moral standards as are human beings. All things, animate and inanimate, are part of the same narrative.

In a moral cosmic order, where all beings are potent moral symbols, what makes humans worthy of their special status is not a unique possession of a soul, reason, or language: it is the display of moral superiority evidenced in their practices. In Jainism, human supremacy and distinction from the nonhuman environment is a matter of degree, not of kind; and, importantly, it is established through ethical behavior. Since nature is a moral theater, and moral perfection is *demonstrable* only through its "acting out" (of which asceticism is the greatest performance), ethical behavior becomes a compelling and potent source of selfhood. Ethics are a resource and represent the primary method through which Jains define and maintain the human domain. *Ahiṃsā* is the quintessential norm of Jain ethics. Its application becomes the way to define human beings at the center of a universe full of similar souls. Voluntary restraint and ardent *ahiṃsā*, in a world characterized by meaningless activity and violence, establish the uniqueness of the human incarnation. Therefore, the ideal moral self in Jainism is created through disengagement and withdrawal from nature.

Jain Ethics in Diaspora

Figure 1. Cartoon on back cover of Jain Spirit, *no. 3 (March–May 2000).*

The ascetic basis of Jainism—its traditional core—is decreasing in significance among a large segment of the Jain immigrant community, particularly the youth. A worldview which emphasizes quiescence and renunciation is commonly seen as "old-fashioned." As one young Jain put it, "Many youth are worried by the outdated and ascetic ideals that are presented to us by the form of Jainism that our parents practice."[16] Others consider ascetic practices to be irrelevant in the modern world. For instance, a Jain college student writes:

> As for the discrepancy between the way of living in the West as opposed to the way that is shown in Jainism, timely changes need to be made. To live both ways of life requires a reinterpretation of the strict rules and outdated practices. For example, it was taught by our ancestors not to eat at night. Today, however, with parents working and daily activities extending into the late evening, it is not possible to eat early. In addition, health standards are safer here than they were hundreds of

years ago and hence, it seems acceptable that we eat late at night. Another example refers to the practice of boiling water before drinking it. Here, there is no need for that. The sanitation department already purifies it before we get it in our homes. Other practices such as these are no longer practical to this day and age.[17]

A participant at the tenth biennial Jaina Convention in Philadelphia reprimands Jainism for its ritualism and calls for a "spiritual awakening," which could be achieved through greater involvement in contemporary social issues:

> Does it not seem like we have been following the rituals blindfolded without ever thinking the spiritual meaning of each Puja or event? Do we want to spend [the] rest of our life in performing robotic rituals or do we want a true spiritual awakening? . . . [Let us] [u]se Ghee Boli funds and donations to help needy Jain families, battered women and children, and set up funds for Jain scholarships on merit basis. Let us bring some missionary spirit in our heart to touch the community.[18]

Another argues that Jainism is ideally suited to help resolve global problems and should be spread beyond the Jain community:

> The coming millennium will be the era of science and technology where only those ethical systems will survive which have high intellectual consistency and internal and external verifiability. The Jaina system passes this test. . . . It has an effective theoretical basis for maintaining environmental equilibrium through observing various vows at all levels. The concepts and practices of equitable distribution and non-absolutist mentality and the like are other tenets which are directly or indirectly influencing the world in a very positive way.[19]

To a great number of Jains in North America, sect, *gaccha* (ascetic lineage), and caste are no longer the significant markers of identity[20] they are in India. There are simply too few Jains to maintain sect endogamy. Instead, in Toronto, all Jain sects are represented in a single Jain center (in Etobicoke) and, unlike in India, intermarriage is common between sects and even outside the faith among strictly vegetarian Hindu families.[21] In North America, *ahiṃsā* and, in particular, its dietary expression in terms of vegetarianism has, for many, become the only non-negotiable characteristic of being Jain.

Results from my research among the Jain community reveal a marked discrepancy between first- and second-generation Jains' understanding of the appropriate application of *ahiṃsā*.[22] For most first-generation Jains, a life of nonviolence (in particular, a vegetarian diet) is intrinsically tied up with ideas of purity and *karma* theory. *Ahiṃsā* is the practice that leads to spiritual liberation. While compassion for all living beings is an important part of their practice, violence (*hiṃsā*) is avoided because it is an obstacle to self-realization. As the Jain scholar Padmanabh Jaini explains:

> . . . for Jainas, [violence] refers primarily to injuring oneself—to behaviour which inhibits the soul's ability to attain *mokṣa*. Thus the killing of animals, for example, is reprehensible not only for the suffering produced in the victims, but even more so because it involves intense passions on the part of the killer, passions which bind him more firmly in the grip of *saṃsāra* [cycle of death and rebirth].[23]

For most Jain youths, however, violence refers principally to harm done to *others*, and *ahiṃsā* is primarily about alleviating the suffering of other living beings. Self-realization is subordinate to this overarching goal.

This sociocentric understanding of Jain ethics has led young Jains to "extend" the practices of nonviolence to areas into which the first generation (and orthodox Jainism more generally) has not ventured. For instance, the use of silk, leather, and dairy products pose an ethical problem for many young Jains that seems not to exist, for the most part, among older, first-generation Jains. The active promotion of a vegan lifestyle (one in which no animal products whatsoever are used or consumed)—while still a minority position—is almost an exclusively second-generation effort.[24] And broad-based interest and activism in environmental issues is also largely a second-generation concern.

The extension of moral consideration into areas previously outside the purview of Jain ethical discourse suggests that an "environmental episteme," as discussed by John Cort in this volume, may in fact be an organic—as opposed to a contrived—development within the diaspora community, however foreign it is to traditional Jain thought. Cort rejects as artificial efforts to graft environmental principles onto Jainism or Jain principles onto environmental practice, and argues instead that a Jain environmental ethic must develop out of a "creative investigation and reinterpretation" of Jain thought and practice. Likewise, Dundas, also in this volume, challenges those efforts that seek to remove Jain teachings from their original social contexts and make them conform to modern values and concerns. In particular, he is suspicious of efforts to redefine as ecocentric a tradition with an "aspiration to withdrawal" from worldly existence at its core. Indeed, such anachronistic efforts at ethical bricolage are superficial and largely unfruitful. But what is occurring within the immigrant community is something entirely different; the emergence of a "Green Jainism" appears to be an entirely natural and involuntarily development within the tradition as it takes root in the sociocultural context of North America.

In the diaspora community, ascetic teachings and modern environmentalism are not being forced into an awkward alliance, uneasily resting on conflicting cosmologies. Instead, asceticism is being deemphasized so that teachings of compassion and nonviolence are no longer anchored to a renunciatory worldview. Jain teachings are being redefined according to a different ethical charter altogether—one in which active engagement in the world is encouraged, one in which asceticism is no longer of central significance.

The traditional Jain concern for the smallest of life-forms (individual *nigodas*) has been transformed into a general concern for "the environment" and is expressed within a discourse of ecology—the need to preserve rain forests, green spaces, clean water. And the traditional Jain concern for animals, as fellow living beings in karmic bondage, is now commonly articulated within an animal rights discourse of rescue and protection. This sociocentric (or "ecological") interpretation of Jain ethics is the governing discourse on Jain web sites and in Jain magazines (such as *Jain Spirit* and *Jiv Daya*) and is widespread at Jain conventions. And significantly, it is also commonly advocated at *pāṭhśāḷās* (schools for religious learning).[25]

Many Jain families in North America send their children to *pāṭhśāḷā*. It is one of the principle ways through which the community hopes to preserve and transmit Jain teachings within the diaspora community. A Toronto Jain Society booklet describes *pāṭhśāḷā* as "a place where children, youth and future generations receive a sound understanding of the Jain philosophy with an equal emphasis on practice of Jainism" (Jain Society of Toronto Souvenir booklet). In addition to learning about Jainism's twenty-four revered teacher-ascetics (Tīrthaṅkaras), rituals and vows, children are also encouraged to apply the principle of nonviolence to remedy perceived social ills. For instance, at a *pāṭhśāḷā* class I attended in October 1999, the teacher asked the class (of about fifteen young people between the ages of twelve and sixteen) to reflect on the following poem:

> Thanksgiving dinner's sad and thankless
> Christmas dinner's dark and blue
> When you stop and try to see it
> From the turkey's point of view
>
> Sunday dinner isn't funny
> Easter Feasts are just bad luck
> When you see it from the viewpoint
> Of a chicken or a duck
>
> Oh how I once loved tuna salad
> Pork and lobsters, lamb chops too
> Till I stopped and looked at dinner
> From the dinner's point of view

The teacher asked, "What principles of Jainism does this poem convey?" The students called out, "*Ahiṃsā*," and "*Anekāntvāda*" (principle of multiple viewpoints). "Precisely," the teacher said. "The poem encourages us to practice nonviolence and to try to see reality from another perspective—in this case from the perspective of the animals."

A cursory examination of contemporary Jain activities within the diaspora reveals how *ahiṃsā* is being established as an ecological ethic. For example:

> A visit to the Toronto Jain Centre discloses this generational and cultural divide: side by side are the Centre's permanent artwork depicting the Jain Tīrthaṅkaras and children's posters made at *pāṭhśāḷās*. The theme of many of the posters concerns human responsibility to the planet. One proclaims: "Don't Smoke; Don't Pollute; Don't Drink Alcohol." Another poster states: "Be Kind to Animals."
>
> Another powerful indicator of this cultural divide can be seen in the dispute over Indian *panjrapoles* (animal sanctuaries). A small but vocal group of young diaspora Jains are highly critical of the *panjrapole* philosophy of noninterference and argue that donating to them "is a

total waste of money."[26] The group's grievance rests with the fact that the *panjrapoles* focus solely on the protection of life and not with the alleviation of animal suffering.

The vast majority of Jain web sites and journals published in the United States, the United Kingdom, and Canada espouse a sociocentric interpretation of Jain ethics.

Young British Jains recently organized two mass rallies in London (on 31 December 1999) to highlight the world's growing environmental crisis and to promote awareness of the vegetarian lifestyle and the current plight of farm animals. Banners at the Vegetarian Rally read, "For a Better and Kinder World: Go Vegetarian." The rationale for vegetarianism is not purity, but social welfare.[27]

"Mahavir Awards"[28] for work in nonviolence were awarded to the following organizations: Compassion in World Farming; PETA (People for the Ethical Treatment of Animals); and the Green Party.[29]

During JAINA Conventions, youth workshops have been organized on the relationship between compassion and ecology.[30]

The Jiv Daya Committee has organized seminars on cruelty-free clothing and cosmetics.[31]

In North America, there has been a strong move by Jain groups to link up with animal rights and environmental organizations. This is evidenced by the high profile that animal rights and ecological issues (as well as their advocates) receive in Jain magazines and at national conventions.

Discussion of Jainism as "inherently" environmental is a common feature in Jain journals, souvenir booklets, and study guides.

An "environment/ecology" council exists as a permanent committee of the JAINA convention.

The reinterpretation of *ahiṃsā*, which I have been describing, reflects more than the inclusion of environmental and animal rights concerns: it reflects a different understanding of "nature" and of the ideal role of human beings *in* nature. A sociocentric interpretation of Jain ethics estranges *ahiṃsā* from its liberation-centric, otherworldly ontology and, in addition, undermines its function in the creation of the Jain ascetic ideal. The differences between the liberation-centric and sociocentric worldviews can be nicely demonstrated by juxtaposing two versions of the same popular tale of the twenty-second "prophet" of Jainism, Tīrthaṅkara Neminath. The first story, told to me in India, reveals the liberation-centric focus of orthodox Jainism; the second is a product of the diaspora community and reveals a distinctly sociocentric focus.

Orthodox, "ascetic" version:
Many years ago there was a handsome young prince called Nemi Kumar. He was to be married to a beautiful princess called Rajimati. On the arranged date the marriage procession started with Nemi Kumar riding the decorated king elephant. All the kings and princes of the Yadav clan joined the procession with their royal regalia and retinue. When the procession was approaching the destination, Nemi Kumar saw that on the side of the road there were large fenced-in areas with cages full of wailing animals and birds. Filled with sympathy and compassion, he asked the elephant driver why the animals and birds were being kept in bondage, The driver informed him that the creatures were collected to be butchered for meat for the large number of guests attending his marriage. Nemi Kumar was filled with despair and a feeling of detachment. He said to the elephant driver, "If I agree to be the cause of the butchering of so many living beings, my life and the one to come will be filled with pain and misery. Therefore, I will not marry. Immediately arrange for the release of all these creatures. Return home to Dwarka." The driver opened the gates of the cages. The animals ran away into the jungle. The driver came back and turned the elephant towards Dwarka. On the way Nemi Kumar took off all the valuables and ornaments on his body and handed them over to the elephant driver. The news spread panic in the marriage procession. All the seniors of the Yadav clan tried to change the mind of Nemi Kumar, but in vain. Nemi Kumar said to them, "As these animals were prisoners in iron cages, we all are prisoners in the cages of karma which is much stronger. See the feeling of joy evident in the animals released from the cages. Know that happiness is in freedom, not in bondage. I want to tread the path of breaking this bondage of karma and embrace eternal bliss." One day, not long afterwards, he stood under an Ashoka tree before many onlookers. There he removed his clothes and pulled out five fistfuls of hair, initiating himself as an ascetic. He spent the next 54 days in deep spiritual practices, meditating and fasting without any attachment to his body. On the fifteenth day of the dark half of the month of Ashvin, while observing a two day fast and meditating, he became

an omniscient. He became the twenty-second *Tirthankara*, known as "*Arhat Neminath*."

When Rajimati [Nemi Kumar's fiancée] recovered from her melancholy, she decided to follow the path taken by Nemi Kumar. When she learned that Nemi Kumar had become an omniscient, she took *diksa* (initiation). She lost herself in penance and other spiritual practices and in the end gained liberation.

Diaspora, "ecological" version:
Many years ago there was a handsome young prince called Nemi Kumar. He was to be married to a beautiful princess called Rajimati. On his wedding day Nemi Kumar led the procession of his family and friends, and his princely retinue, towards Princess Rajimati's palace. Everyone was in a festive mood. There was music in the air. The Prince was sitting calmly in his chariot, which his charioteer was driving. Suddenly Prince Nemi Kumar heard animal noises which got louder as they got nearer. They soon saw where the noises were coming from. Prince Nemi Kumar asked the procession to stop and listen. Hundreds of animals and birds were packed tightly in cages. There were fish in large tanks. The animals seemed frightened and restless. Their eyes were pleading. The Prince asked his friends why these animals and birds were captured. He was told they were for his wedding feast. This saddened the Prince, who was very kind and sensitive. The frightened sheep seemed to say, "We will be slaughtered for this prince's feast." A beautiful deer had his eyes full of tears, as if he were pleading, "I don't want to be killed, I want to go back to the forest and roam free." Beautiful green parrots were flying here and there in their cages trying to find a way out. A wise bull seemed to be saying, "These men are cruel. They cry when their children die, but how can they kill our children? Why can't they eat only plants and fruits, as we do? How can they claim to be superior to us when they kill us all the time?" The kind Prince could bear it no longer. His heart was crying at the pain and fear the poor animals were suffering. He climbed down from his chariot and walked towards the cages. The animals quietened down, seeing such a stately but kind and loving figure walking towards them. They knew that they need no longer be frightened. The Prince opened the cages, and let the animals and birds out. He told his men to return the fish to the sea without harming them. The birds flew out happy and free. The animals ran into the forest. They all seemed to be thanking the Prince for saving them. Just then King Ugrasen, the Princess's father, came to meet the Prince. He saw the Prince releasing the animals and asked, "Why have you released these animals, O Prince?" The Prince replied,

> "How can we rejoice when so many animals are suffering? *How can we humans feast on these innocent animals and birds we are meant to protect?* What use is happiness if it is built on the suffering of so many? With this the Prince turned his chariot and went back. The wedding was called off. After some time, the Prince became a monk. Princess Rajimati followed in his footsteps and became a nun. The Prince Nemi Kumar was none other than the 22nd Tirthankar Bhagvan Neminath.[32]

What is lost in the diaspora version is what is most meaningful to Jains in India, namely, the ascetic values of detachment and renunciation. First, it estranges asceticism, and makes its connection with *ahiṃsā*, or nonviolence, rather puzzling. Second, although it retains the centrality of the doctrine of *ahiṃsā*, it does so with an emphasis on the importance of suffering. A preoccupation with the avoidance of suffering betrays a this-worldly orientation, in that it assumes suffering is so wretched because it is so meaningless.

In the diaspora version, Nemi Kumar asks, "How can we humans feast on these innocent animals and birds we are meant to protect?" This idea of "protecting the innocent" implies that it is only humans who are endowed with moral judgment, which humans should use to take care of "instinctual" beasts. But a central tenet of all Indian religious traditions is that the universe is a moral order where all things are endowed with moral status.[33] Suffering may be abominable, but it is never meaningless. Moral law, and not the mechanical forces of nature, govern and control the world and all its processes. This is distinct from a "nature/ culture" worldview, which considers morality a human peculiarity in a passive, innocent nature and which, in turn, assumes an ethic of active support to be a uniquely human responsibility. Crusades to rid the world of suffering presuppose the human ability to do so; only humans with their "humanity" or "civilization" can bring morality to a profane, arbitrary "nature."

We conclude from the diaspora version of the story that Prince Nemi Kumar behaved righteously because he saved the lives of the animals. His reasons for renouncing the world and becoming a monk are not at all clear. If the alleviation of physical suffering was his motive, why not continue to do so? Almost without exception, Jain stories end with the protagonist renouncing the world and embarking on the ascetic path. We do not learn about all the further austerities Nemi Kumar deliberately put himself through on his path of asceticism, perhaps because that would be difficult to reconcile with a story first and foremost about the avoidance of suffering. The original version, by contrast, does not stress the avoidance of suffering. Suffering has never been the primary concern of the Jain ethical system, precisely because its existence can be rationalized within a moral cosmic order. The orthodox understanding of *ahiṃsā* is primarily concerned with the *avoidance* of behavior that inhibits the soul's ability to attain *mokṣa*, not an injunction to alleviate suffering. Charles Taylor argues that modern Westerners place an exceptional importance on avoiding suffering, far more now than even just a few centuries ago, and, significantly, he attributes this to a decline in the West of the whole notion of a moral cosmic order, which gave misfortune "meaning."

The orthodox version of the story places stress on detachment because *attachment* is the root of violence and the source of all bondage. The killing of animals is an extreme form of violence arising from attachment, but Nemi Kumar realizes that social life itself is inherently violent. Marriage, for example, as an attachment is also a form of violence. Asceticism, as a "stepping out" of society, becomes logical: the ascetic path is the best means to ensure a life of detachment. And while compassion is a feature of the original story, it is not understood as emotional vicarious suffering; instead, it is presented as respect for all living beings as equal souls. Every soul is entrapped in worldly bondage and will one day have to break those bonds if liberation is to be attained. Compassion means recognizing that all living beings are essentially similar, that all deserve respect, and that none should be injured. It means not interfering in another's spiritual journey.

Orthodox Jainism is intrinsically otherworldly. Although it espouses a powerful ethos of respect and compassion for all living beings, it is not an ethic of social activisim. The Jain ethic of nonharm *is* a powerful ecology in itself, but its teachings are not designed to remedy social ills so much as to escape them. Orthodox Jain ethics reveal a perception of the world as inherently corrupt and in need of transcendence, and this perception leads to renunciation and to the desire to help individuals out of *saṃsāra*, not to active social involvement.

In addition, traditional Jainism problematizes "nature." Nature, or worldly existence, is the perennial obstacle to transcendence. Infused with a myriad of life-forms, it is a minefield of destruction and unavoidable violence. The proper relationship with nature is one of detachment, withdrawal, and vigilance. The understanding of "nature" that is being fostered in the North American context is profoundly different. In a sociocentric ecological worldview, the ideal role for human beings is to become actively engaged with nature—to develop an ecology of love and enchantment in order to preserve it. Most Jains of the diaspora are at ease with this ethos and readily identify Jain compassion and nonviolence with it. This view is, however, as alien to traditional Jainism as is the modernist view of nature as dead matter. Love of and attachment to nature would be *mithyādṛṣṭi* (deluded belief) and would be antithetical to the path of purification.

Jainism, as it is being constituted in North America, reflects its new roots. Its traditional ethics of nonviolence, self-control, and renunciation are being divorced from the traditional ontology and redefined in ways that better reflect the concerns of the modern diaspora community. There has never been a monolithic Jainism; different and competing interpretations have always existed. Ecocentric diaspora Jainism may simply be one of the tradition's newest expressions. It does, however, mark a radical departure from other forms in its rejection of the renunciatory ethic that has always been the incontestable and abiding core of Jainism.[34]

In spite of their divergences, the ethics of the orthodox and the diaspora communities both stress the centrality of "nature" in the constitution of the moral self. Both emphasize respect for individual living beings as subjects, and both consider human relations with the nonhuman world to be of central importance. For both, "nature" is the moral theater within which one's ethical being is established—a fact that may very well be one of the most important features of Jain ethical life.

Notes

I am indebted to the Jains of Toronto for their tremendous kindness and generosity. The community's sincere and active involvement with the research has contributed so much to the process. Thank you. In addition, I am grateful for the financial and academic support provided by the Rockefeller Postdoctoral Fellowship in the Humanities between 1999 and 2000.

1. This chapter is based on research I have been pursuing on the process of identity construction among the Jain immigrant community in North America. In particular, I am looking at the strategies of Jain identity formation and maintenance within multiethnic, religiously pluralistic metropolitan Toronto, home to nearly two-thirds of the Jain community in Canada. Although I believe the phenomena that I am describing here is true for North American Jainism in general, the focus of my research has been the Toronto Jain community.

2. It should be noted that an order of *samans*, or "semi-ascetics" (those that have not taken the full vows of asceticism and as such are permitted to travel outside of India), do visit North America for several months at a time. Although they purport to speak on behalf of Jainism in general, they are, in effect, representatives of the Terāpanthī Śvetāmbara sect and primarily seek to raise awareness about this sect. Their impact within the larger Jain community is limited, especially since the majority of North American Jains are not associated with Terapanthism.

Harold Coward emphasized the pivotal role the guru plays among immigrant Hindu communities in Alberta (David J. Goa, Harold G. Coward, and Ronald Neufeldt, "Hindus in Alberta: A Study in Religious Continuity and Change," *Canadian Ethnic Studies* 16, no. 1 [1984]: 96–113). These communities have argued that in the Canadian setting, the guru's role takes on even more significance than in the motherland. In addition to transmitting and preserving cultural knowledge, the guru acts as a link between the old and new homes and serves as a locus of adaptation to Canadian life. Since there are no Jain ascetics outside of India, the diaspora community lacks these central community builders.

3. Lawrence Babb, *Absent Lord: Ascetics and Kings in a Jain Ritual Culture* (Berkeley: University of California Press, 1996).

4. See *In the Further Soil: A Social History of Indo-Canadians in Ontario*, ed. Milton Israel (Richmond Hill, Ontario: Organisation for the Promotion of Indian Culture, 1994).

5. Paul Dundas, *The Jains* (London and New York: Routledge, 1992).

6. Jain Centre of Toronto, 1998.

7. Dundas, *The Jains*, 3.

8. Marcus Banks, "Orthodoxy and Dissent: Varieties of Religious Belief among Immigrant Gujarati Jains in Britain," in *The Assembly of Listeners: Jains in Society*, ed. Michael Carrithers and Caroline Humphrey (Cambridge: Cambridge University Press, 1991).

9. Dundas, *The Jains*, 232–33.

10. *The Assembly of Listeners: Jains in Society*, ed. Michael Carrithers and Caroline Humphrey (Cambridge: Cambridge University Press, 1991).

11. James Laidlaw, *Riches and Renunciation: Religion, Economy and Society among the Jains* (Oxford: Clarendon Press, 1995)

12. Dundas, *The Jains*, 36–37.

13. The Jain concern with the tiniest of life-forms—*nigoda*s—has long been a source of mockery (ridiculed as a preoccupation with "insects" or with "bacteria"), puzzlement, and even frustration among those who would like to see greater Jain involvement in broader social issues. The Jain concern with the simplest life-forms must be understood within the context of its understanding of nature as a "moral theater." Rather than being trivial, it is the harm done to these small life-forms (each endowed with a soul) that is the primary cause of our karmic bondage

14. For a detailed and erudite discussion of the five categories of life in Jainism, see Kristi L. Wiley, "The Nature of Nature: Jain Perspectives on the Natural World," in this volume.

15. Wiley, in this volume, points out that this path is not necessarily a unilineal trajectory, in which the soul must pass through a fixed number of stages on the way to liberation. Souls may progress, regress, skip stages, or remain at the same level for many births.

16. Resma Modi, "Living a Jain Way of Life in the Western Environment" (youth essay contest entry, group 2, college age, Seventh Biennial Jaina Convention, Pittsburgh, 1993).

17. Avani Doshi, "The Future of Jainism in the West" (youth essay contest entry, group 2, college age, Seventh Biennial Jaina Convention, Pittsburg, 1993). This public-health interpretation of traditional ascetic practices ignores the original motivation underlying the restrictions, namely, the avoidance of harming minute living beings—in this case, fire-bodied and water-bodied beings.

18. Pravin Shah, "Jainism in the Next Millennium," Philadelphia Jaina Convention Souvenir booklet, 1999, 49.

19. N. L. Jain, "Promotion of Jainism in the Next Millennium," Philadelphia Jaina Convention Souvenir booklet, 1999, 139.

20. On a survey/questionnaire that I have recently distributed among the community, a number of people expressed reservation in answering the question: "To which sect do you belong?" I was told that answering this seemed counterproductive to their efforts to eradicate sect-consciousness.

21. It is not my intention to portray an overly romantic picture of the North American Jain community. The majority of North American Jains are Gujarati-speaking Deravāsīs and, because of their numbers, they have the most influence in matters affecting the community—often to the frustration of other sects. Nevertheless, among the younger people, sect is a far less important consideration than it is for their parents.

22. My research primarily consisted of informal meetings and interviews with Jains from all parts of North America, though the focus was on Toronto Jains. In addition, I conducted a survey within the Toronto Jain community.

23. Padmanabh Jaini, *Ahimsa*, The 1990 Inaugural Roop Lal Jain Lecture (Toronto: Centre for South Asian Studies, University of Toronto, 1990), 167.

24. Although veganism is practised by only a small number of Jains, the ethical

issues upon which it is based have a strong appeal (and are widely endorsed, if not practised) among younger Jains.

25. This, however, largely depends on whether the teacher is a first- or second-generation Canadian or American Jain.

26. *Jain Digest* 16, no. 1 (2000): 12.

27. See *Jain Spirit*, no. 3 (March–May 2000).

28. Lord Mahāvīra is the revered twenty-fourth and final Tīrthaṅkara of our present age.

29. See *Jain Spirit*, no. 3 (March–May 2000).

30. For example, see the Ninth Jaina Convention Souvenir booklet, 1997, 123.

31. Ibid.

32. Vinod Kapashi et al., *Text Book of Jainism, Level 1* (Middlesex: The Institute of Jainology, 1994), 16–17. Emphasis added.

33. Saral Jhingran, *Aspects of Hindu Morality* (Delhi: Motilal Banarsidass, 1989), 33.

34. See Paul Dundas, "The Limits of a Jain Environmental Ethic," in this volume.

Appendix: The Jain Declaration on Nature

L. M. SINGHVI

The Jain tradition which enthroned the philosophy of ecological harmony and non-violence as its lodestar flourished for centuries side-by-side with other schools of thought in ancient India. It formed a vital part of the mainstream of ancient Indian life, contributing greatly to its philosophical, artistic and political heritage. During certain periods of Indian history, many ruling elites as well as large sections of the population were Jains, followers of the *Jinas* (Spiritual Victors).

The ecological philosophy of Jainism which flows from its spiritual quest has always been central to its ethics, aesthetics, art, literature, economics and politics. It is represented in all its glory by the 24 *Jinas* or *Tirthankaras* (Path-finders) of this era whose example and teachings have been its living legacy through the millennia.

Although the ten million Jains estimated to live in modern India constitute a tiny fraction of its population, the message and motifs of the Jain perspective, its reverence for life in all forms, its commitment to the progress of human civilization and to the preservation of the natural environment continue to have a profound and pervasive influence on Indian life and outlook.

In the twentieth century, the most vibrant and illustrious example of Jain influence was that of Mahatma Gandhi, acclaimed as the Father of the Nation. Gandhi's friend, Shrimad Rajchandra, was a Jain. The two great men corresponded, until Rajchandra's death, on issues

Reprinted by permission of L. M. Singhvi.

of faith and ethics. The central Jain teaching of *ahimsa* (non-violence) was the guiding principle of Gandhi's civil disobedience in the cause of freedom and social equality. His ecological philosophy found apt expression in his observation that the greatest work of humanity could not match the smallest wonder of nature.

I. The Jain Teachings

1. *Ahimsa* (non-violence)

The Jain ecological philosophy is virtually synonymous with the principle of *ahimsa* (non-violence) which runs through the Jain tradition like a golden thread.

> "*Ahimsa parmo dharmah*" (Non-violence is the supreme religion)

Mahavira, the 24th and last *Tirthankara* (Path-finder) of this era, who lived 2500 years ago in north India consolidated the basic Jain teachings of peace, harmony and renunciation taught two centuries earlier by the *Tirthankara* Parshva, and for thousands of years previously by the 22 other *Tirthankaras* of this era, beginning with Adinatha Rishabha. Mahavira threw new light on the perennial quest of the soul with the truth and discipline of *ahimsa*. He said:

> "There is nothing so small and subtle as the atom nor any element so vast as space. Similarly, there is no quality of soul more subtle than non-violence and no virtue of spirit greater than reverence for life."

Ahimsa is a principle that Jains teach and practice not only towards human beings but towards all nature. It is an unequivocal teaching that is at once ancient and contemporary. The scriptures tell us:

> "All the *Arhats* (Venerable Ones) of the past, present and future discourse, counsel, proclaim, propound and prescribe thus in unison: Do not injure, abuse, oppress, enslave, insult, torment, torture or kill any creature of living being."

In this strife-torn world of hatred and hostilities, aggression and aggrandizement, and of unscrupulous and unbridled exploitation and consumerism, the Jain perspective finds the evil of violence writ large.

The teaching of *ahimsa* refers not only to wars and visible physical acts of violence but to violence in the hearts and minds of human beings, their lack of concern and compassion for their fellow humans and for the natural world. Ancient Jain texts explain that violence (*himsa*) is not defined by actual harm, for this may be unintentional. It is the intention to harm, the absence of compassion, that makes action violent. Without violent thought there could be no violent actions. When violence enters our thoughts, we remember *Tirthankara* Mahavira's words:

> "You are what you intend to hit, injure, insult, torment, persecute, torture, enslave or kill."

2. *Parasparopagraho jivinam* (interdependence)

Mahavira proclaimed a profound truth for all times to come when he said:

> "One who neglects or disregards the existence of earth, air, fire, water and vegetation disregards his own existence which is entwined with them."

Jain cosmology recognizes the fundamental natural phenomenon of symbiosis or mutual dependence, which forms the basis of modern-day science of ecology. It is relevant to recall that the term 'ecology' was coined in the latter half of the nineteenth century from the Greek word *oikos*, meaning 'home', a place to which one returns. Ecology is the branch of biology which deals with the relationships of organisms to their surroundings and to other organisms.

The ancient Jain scriptural aphorism *Parasparopagraho jivinam* (All life is bound together by mutual support and interdependence) is refreshingly contemporary in its premise and perspective. It defines the scope of modern ecology while extending it further to a more spacious 'home.' It means that all aspects of nature belong together and are bound in a physical as well as a metaphysical relationship. Life is viewed as a gift of togetherness, accommodation and assistance in a universe teeming with interdependent constituents.

3. *Anekantavada* (the doctrine of manifold aspects)

The concept of universal interdependence underpins the Jain theory of knowledge, known as *anekantavada* or the doctrine of manifold aspects. *Anekantavada* describes the world as a multifaceted, ever-changing reality with an infinity of viewpoints depending on the time, place, nature and state of the one who is the viewer and that which is viewed.

This leads to the doctrine of *syadvada* or relativity, which states that truth is relative to different viewpoints (*nayas*). What is true from one point of view is open to question from another. Absolute truth cannot be grasped from any particular viewpoint alone because absolute truth is the sum total of all the different viewpoints that make up the universe.

Because it is rooted in the doctrines of *anekantavada* and *syadvada*, Jainism does not look upon the universe from an anthropocentric, ethnocentric or egocentric viewpoint. It takes into account the viewpoints of other species, other communities and nations and other human beings.

4. *Samyaktva* (equanimity)

The discipline of non-violence, the recognition of universal interdependence and the logic of the doctrine of manifold aspects, leads inexorably to the avoidance of dogmatic, intolerant, inflexible, aggressive, harmful and unilateral attitudes towards the world around. It inspires the personal quest of every Jain for *samyaktva* (equanimity) towards both *jiva* (animate beings) and *ajiva* (inanimate substances and objects). It encourages an attitude of give and take and of live and let live. It offers a pragmatic peace plan based, not on the domination of nature, nations or other people, but on an equanimity of mind devoted to the preservation of the balance of the universe.

5. *Jiva-daya* (compassion, empathy and charity)

Although the term '*ahimsa*' is stated in the negative (*a*=non, *himsa*= violence), it is rooted in a host of positive aims and actions which have great relevance to contemporary environmental concerns.

Ahimsa is an aspect of *daya* (compassion, empathy and charity), described by a great Jain teacher as "the beneficent mother of all beings" and "the elixir for those who wander in suffering through the ocean of successive rebirths."

Jiva-daya means caring for and sharing with all living beings, tending, protecting and serving them. It entails universal friendliness (*maitri*), universal forgiveness (*kshama*) and universal fearlessness (*abhaya*).

Jains, whether monks, nuns or householders, therefore, affirm prayerfully and sincerely, that their heart is filled with forgiveness for all living beings and that they have sought and received the forgiveness of all beings, that they crave the friendship of all beings, that all beings give them their friendship and that there is not the slightest feeling of alienation or enmity in their heart for anyone or anything. They also pray that forgiveness and friendliness may reign through the world and that all living beings may cherish each other.

II. Jain Cosmology

Jains do not acknowledge an intelligent first cause as the creator of the universe. The Jain theory is that the universe has no beginning or end. It is traced to *jiva* and *ajiva*, the two everlasting, uncreated, independent and co-existing categories. Consciousness is *jiva*. That which has no consciousness is *ajiva*.

There are five substances of *ajiva:*

- *dharma* - the medium of motion
- *adharma* - the medium of rest
- *akasha* - space
- *pudgala* - matter
- *kala* - time

Pudgala (matter) has form and consists of individual atoms (*paramanu*) and conglomerates of atoms (*skandha*) which can be seen, heard, smelled, tasted and/or touched. According to Jains, energy, or the phenomena of sound, darkness, shade, heat, light and the like, is produced by conglomerates of atoms.

The *jiva* (soul) has no form but, during its worldly career, it is vested with a body and becomes subject to an inflow of karmic 'dust' (*asravas*). These are the subtle material particles that are drawn to a soul because of its worldly activities. The *asravas* bind the soul to the physical world until they have brought about their karmic result when they fall away 'like ripe fruit' by which time other actions have drawn more *asravas* to the soul.

With the exception of the *Arihantas* (the Ever-Perfect) and the

Siddhas (the Liberated), who have dispelled the passions which provide the 'glue' for the *asravas*, all souls are in karmic bondage to the universe. They go through a continuous cycle of death and rebirth in a personal evolution that can lead at last to *moksha* (eternal release). In this cycle there are countless souls at different stages of their personal evolution: earth-bodies, water-bodies, fire-bodies, air-bodies, vegetable-bodies, and mobile bodies ranging from bacteria, insects, worms, birds and larger animals to human beings, infernal beings and celestial beings.

The Jain evolutionary theory is based on a grading of the physical bodies containing souls according to the degree of sensory perception. All souls are equal but are bound by varying amounts of *asravas* (karmic particles) which is reflected in the type of body they inhabit. The lowest form of physical body has only the sense of touch. Trees and vegetation have the sense of touch and are therefore able to experience pleasure and pain, and have souls. Mahavira taught that only the one who understood the grave demerit and detriment caused by destruction of plants and trees understood the meaning and merit of reverence for nature. Even metals and stones might have life in them and should not be dealt with recklessly.

Above the single-sense *jivas* are micro-organisms and small animals with two, three or four senses. Higher in the order are the *jivas* with five senses. The highest grade of animals and human beings also possess rationality and intuition (*manas*). As a highly evolved form of life, human beings have a great moral responsibility in their mutual dealings and in their relationship with the rest of the universe.

It is this conception of life and its eternal coherence, in which human beings have an inescapable ethical responsibility, that made the Jain tradition a cradle for the creed of environmental protection and harmony.

III. The Jain Code of Conduct

1. The five *vratas* (vows)

The five *vratas* (vows) in the Jain code of conduct are:

- non-violence in thought, word and deed,
- to seek and speak the truth,

- to behave honestly and never take anything by force or theft,
- to practice restraint and chastity in thought, word and deed,
- to practice non-acquisitiveness.

The vow of *ahimsa* is the first and pivotal vow. The other vows may be viewed as aspects of *ahimsa* which together form an integrated code of conduct in the individual's quest for equanimity and the three jewels (*ratna-traya*) of right faith, right knowledge and right conduct.

The vows are undertaken at an austere and exacting level by the monks and nuns and are then called *maha-vratas* (great vows). They are undertaken at a more moderate and flexible level by householders and called the *anu-vratas* ('atomic' or basic vows).

Underlying the Jain code of conduct is the emphatic assertion of individual responsibility towards one and all. Indeed, the entire universe is the forum of one's own conscience. The code is profoundly ecological in its secular thrust and its practical consequences.

2. Kindness to animals

The transgressions against the vow of non-violence include all forms of cruelty to animals and human beings. Many centuries ago, Jains condemned as evil the common practice of animal sacrifice to the gods. It is generally forbidden to keep animals in captivity, to whip, mutilate or overload them or to deprive them of adequate food and drink. The injunction is modified in respect of domestic animals to the extent that they may be roped or even whipped occasionally but always mercifully with due consideration and without anger.

3. Vegetarianism

Except for allowing themselves a judicious use of one-sensed life in the form of vegetables, Jains would not consciously take any life for food or sport. As a community they are strict vegetarians, consuming neither meat, fish nor eggs. They confine themselves to vegetable and milk products.

4. Self-restraint and the avoidance of waste

By taking the basic vows, the Jain laity endeavor to live a life of moderation and restraint and to practice a measure of abstinence and austerity. They must not procreate indiscriminately lest they overburden the universe and its resources. Regular periods of fasting for self-purification are encouraged.

In their use of the earth's resources Jains take their cues from "the bee [that] sucks honey in the blossoms of a tree without hurting the blossom and strengthens itself." Wants should be reduced, desires curbed and consumption levels kept within reasonable limits. Using any resource beyond one's needs and misuse of any part of nature is considered a form of theft. Indeed, the Jain faith goes one radical step further and declares unequivocally that waste and creating pollution are acts of violence.

5. Charity

Accumulation of possessions and enjoyment for personal ends should be minimized. Giving charitable donations and one's time for community projects generously is a part of a Jain householder's obligations. That explains why the Jain temples and pilgrimage centers are well-endowed and well-managed. It is this sense of social obligation born out of religious teachings that has led the Jains to found and maintain innumerable schools, colleges, hospitals, clinics, lodging houses, hostels, orphanages, relief and rehabilitation camps for the handicapped, old, sick and disadvantaged as well as hospitals for ailing birds and animals. Wealthy individuals are advised to recognize that beyond a certain point their wealth is superfluous to their needs and that they should manage the surplus as trustees for social benefit.

* * *

The five fundamental teachings of Jainism and the five-fold Jain code of conduct outlined in this declaration are deeply rooted in its living ethos in unbroken continuity across the centuries. They offer the world today a time-tested anchor of moral imperatives and a viable route plan for humanity's common pilgrimage for holistic environmental protection, peace and harmony in the universe.

Select Bibliography

Ācārāṅga Sūtra (Ayaro). New Delhi: Today and Tomorrow's Printers and Publishers, 1981.

Ādipurāṇa of Jinasena. Sanskrit text with Hindi, translated by Pannalal Jain. 2 vols. Varanasi: Bhāratīya Jñānapīṭha, 1963–1965.

Akalaṅka. *Sanmati Tarka.* Edited by S. Sanhhavi and B. Doshi. Ahmedabad: Gujarat Paratattva Mandira Granthavali, 1924–1931.

Anuvibha Reporter (Journal of Anuvrat Global Organization, Jaipur). Vol. 3, no. 1., 1997.

Babb, Lawrence A. *Absent Lord: Ascetics and Kings in a Jain Ritual Culture.* Berkeley and Los Angeles: University of California Press, 1996.

Banks, Marcus. *Organizing Jainism in India and England.* Oxford: Clarendon Press, 1992.

———. "Orthodoxy and Dissent: Varieties of Religious Belief among Immigrant Gujarati Jains in Britain." In *The Assembly of Listeners: Jains in Society,* edited by Michael Carrithers and Caroline Humphrey. Cambridge: Cambridge University Press, 1991.

———. "Representing the Bodies of the Jains." In *Rethinking Visual Anthropology,* edited by Marcus Banks and Howard Morphy, 216–39. New Haven: Yale University Press, 1997.

Bhadrankaravijaya, Muni, and Muni Kalyanaprabhavavijaya, eds. *Sraddha-Pratikrama Sutra (Prabodha Tika).* Vol. 2. Bombay: Jain Sahitya Vikasa Mandala, n.d.

Bhagavatī Sūtra. Translated by K. C. Lalwani. 3 vols. Calcutta: Jain Bhawan, 1973–1980.

Bhargava, D. N. "Pathological Impact of [sic] Environment of Professions Prohibited by Jain Acaryas." In *Medieval Jainism: Culture and Environment,* edited by P. S. Jain and R. M. Lodha, 103–8. New Delhi: Ashish Publishing House, 1990.

Bose, Nirmal Kumar, ed. *Selections from Gandhi.* Ahmedabad: Navajivan Publishing House, 1957.

Bourdieu, Pierre. *Outline of a Theory of Practice.* Translated by Richard Nice. Cambridge: Cambridge University Press, 1977.

Carrithers, Michael, and Caroline Humphrey, eds. *The Assembly of Listeners: Jains in Society.* Cambridge: Cambridge University Press, 1991.

Chapple, Christopher Key. "Jainism and Buddhism." In *A Companion to Environmental Philosophy,* edited by Dale Jamieson, 52–66. Malden, Mass.: Blackwell Publishers, 2001.

_______. "Jainism and Nonviolence." In *Subverting Hatred: The Challenge of Nonviolence in Religious Traditions,* edited by Daniel L. Smith-Christopher. Cambridge, Mass.: Boston Research Center for the 21st Century, 1998.

_______. "Life Force in Jainism and Yoga." In *The Meaning of Life in the World Religions,* edited by Joseph Runzo and Nancy M. Martin, 137–50. Oxford: Oneworld, 2000.

_______. *Nonviolence to Animals, Earth and Self in Asian Traditions.* Albany: State University of New York Press, 1993.

_______. "Toward an Indigenous Indian Environmentalism." In *Purifying the Earthly Body of God: Religion and Ecology in Hindu India,* edited by Lance E. Nelson. Albany: State University of New York Press, 1998.

Coomaraswamy, Ananda K. *The Transformation of Nature in Art.* Cambridge, Mass.: Harvard University Press, 1934.

Cort, John E. "Intellectual *Ahiṃsā* Revisited: Jain Tolerance and Intolerance of Others." *Philosophy East and West* 50, no. 3 (2000): 324–47.

_______. "Jain Questions and Answers: Who Is God and How Is He Worshiped?" In *Religions of India in Practice,* edited by Donald S. Lopez, Jr., 598–608. Princeton, N.J.: Princeton University Press, 1995.

_______. *Jains in the World: Religious Values and Ideology in India.* Oxford and New York: Oxford University Press, 2001.

_______. "Liberation and Wellbeing: A Study of Svetambara Muripujak Jains of North Gujarat." Ph.D. diss., Harvard University, 1989.

_______, ed. *Open Boundaries: Jain Communities and Cultures in Indian History.* Albany: State University of New York Press, 1998.

_______. "Recent Fieldwork Studies of the Contemporary Jains." *Religious Studies Review* 23 (1997): 103–11.

_______. "The Rite of Veneration of Jina Images." In *Religions of India in Practice,* edited by Donald S. Lopez, Jr., 326–32. Princeton, N.J.: Princeton University Press, 1995.

Coward, Harold. "New Theology on Population, Consumption, and Ecology." *Journal of the American Academy of Religion* 65 (1997): 259–73.

Coward, Harold, and D. Goa. "Religious Experience of the South Asia

Diaspora in Canada." In *The South Asian Diaspora in Canada: Six Essays*, edited by Milton Israel, 73–86. Toronto: Multicultural History Society of Ontario, 1987.

Deleu, Jozef. *Viyāhapannatti (Bhagavaī): The Fifth Anga of the Jaina Canon*. Brugge: De Tempel, Tempelhof, 1970.

Dhruva, A. B., ed. *Malliṣeṇa, Syada-manjari*. Bombay: Bombay University, 1933.

Doshi, Avani. "The Future of Jainism in the West." Youth essay contest entry, in Jaina Convention Souvenir, Pittsburg, 1993.

Dregson, Alan, and Yuichi Inoue, eds. *The Deep Ecology Movement: An Introductory Anthology*. Berkeley: North Atlantic Books, 1995.

Dundas, Paul. *The Jains*. London: Routledge, 1992.

_______. *The Meat at the Wedding Feast: Krsna, Vegetarianism and a Jain Dispute*. The 1997 Roop Lal Jain Lecture. Toronto: Centre for South Asian Studies, University of Toronto, 1997.

_______. "Recent Research on Jainism." *Religious Studies Review* 23 (1997): 113–19.

Gaeffke, Peter. "The Concept of Nature in the Literature of Bengali, Hindi and Urdu around 1900." *Studien zur Indologie und Iranistik* 10 (1988): 79–101.

Glasnap, Helmuth von. *The Doctrine of Karma in Jain Philosophy*. Translated by G. Barry Gifford. Bombay: Bai Vijibhai Jivanlal Pannalal Charity Fund, 1992.

Gommaṭasāra Jīvakāṇḍa. 2 vols. Jñānapīṭha Mūrtidevī Jaina Granthamālā, Prakrit Grantha, nos. 14–15. New Delhi: Bhāratīya Jñānapīṭha, 1978–1979.

Gommaṭasāra Jīvakāṇḍa, translated by J. L. Jaini. Sacred Books of the Jainas, vol. 5. 1927. Reprint, New Delhi: Today and Tomorrow's Printers and Publishers, 1990.

Granoff, Phyllis. "The Violence of Non-violence: A Study of some Jain Responses to Non-Jain Religious Practices." *Journal of the International Association of Buddhist Studies* 15 (1992): 1–45.

Guha, Ramachandra, and Juan Martinez-Alier. *Varieties of Environmentalism: Essays North and South*. London: Earthscan, 1997.

Hemacandra. *Triṣaṣṭiśalākāpuruṣacaritra: The Lives of Sixty-three Illustrious Persons*. Translated by Helen Johnson. 6 vols. Baroda: Oriental Institute, 1962.

Holmstorm, Savitri. "Towards a Police of Renunciation: Jain Women and Asceticism in Rajasthan." M.A. thesis, University of Edinburgh, 1988.

Israel, Milton. *In Further Soil: A Social History of Indo-Canadians in Ontario*. Richmond Hill, Ontario: Organisation for the Promotion of Indian Culture, 1994.

———. Introduction to *The South Asian Diaspora in Canada: Six Essays*, edited by Milton Israel, 9–14. Toronto: Multicultural History Society of Ontario, 1987.

Jain, N. L. *Biology in Jaina Treatise on Reals (Biology in Tattvārtha-Sūtra): English Translation with Notes on Chapter Two of Tattvārtha-Rājavārtika of Akalaṅka (Royal Semi-Aphorismic Explanatory on Reals) on Tattvārtha-Sūtra (Treatise on Reals) by Ācārya Umāsvāti.* Varanasi: Pārśvanātha Vidyāpiṭha; Chennai: Śri Digambar Jain Samāj, 1999.

———. "Promotion of Jainism in the Next Millenium." In Philadelphia Jaina Convention Souvenir booklet, 1999.

Jaina Sūtras. Part 1. *The Ākārāṅga Sūtra. The Kalpa Sūtra.* Translated by Hermann Jacobi. 1884. Reprint, Delhi: Motilal Banarsidass, 1989.

Jaina Sūtras. Part 2. *The Uttarādhyayana Sūtra. The Sūtrakṛtāṅga Sūtra.* Translated by Hermann Jacobi. 1895. Reprint, Delhi: Motilal Banarsidass, 1989.

Jaini, Padmanabh S. "Ahimsa." Inaugural Roop Lal Jain Lecture. Toronto: Centre for South Asian Studies, University of Toronto, 1990.

———. "Bhavyatva and Abhavyatva: A Jain Doctrine of Predestination." In *Mahāvīra and His Teachings*, edited by A. N. Upadhye et al., 95–111. Bombay: Bhagavān Mahāvīra 2500th Nirvāna Mahotsava Samiti, 1977.

———. *Collected Papers on Jain Studies.* Delhi: Motilal Banarsidass, 2000.

———. "Fear of Food? Jain Attitudes of Eating." In *Jain Studies in Honour of Jozef Deleu*, edited by Rudy Smet and Kenji Watanabe, 339–53. Tokyo: Hon-no-Tomosha, 1993.

———. "From Nigoda to Mokṣa: The Story of Marudevī." In *Proceedings from the International Conference on Jainism and Early Buddhism, University of Lund*, June 4–7, 1998.

———. *Gender and Salvation: Jaina Debates on the Spiritual Liberation of Women.* Berkeley and Los Angeles: University of California Press, 1991.

———. "Indian Perspectives on the Spirituality of Animals." In *Buddhist Philosophy and Culture: Essays in Honour of N. A. Jayawikrema*, edited by David J. Kalupahana and W. G. Weerarante, 169–78. Colombo: N. A. Jayawickrema Felicitation Volume Committee, 1987.

———. *The Jaina Path of Purification.* Berkeley and Los Angeles: University of California Press, 1979.

Johnson, W. J. "The Religious Function of Jaina Philosophy: *Anekāntavāda* Reconsidered." *Religion* 25, no 1 (1995): 41.

Kapadia, H. R., ed. *Haribhadra: Anekāntajayapatāka.* Vol. 2. Baroda: Oriental Institute, 1947.

Kelting, Mary Whitney. "Hearing the Voices of the Śrāvikā: Ritual and Song in Jain Laywomen's Beliefs and Practice." Ph.D. diss., University of Wisconsin, 1996.

Koller, John M., and Patricia Koller. *Asian Philosophies.* Upper Saddle River, N.J.: Prentice Hall, 1998.

Kraft, Kenneth. "The Ecological Implications of Karma Theory." In *Purifying the Earthly Body of God: Religion and Ecology in Hindu India,* edited by Lance E. Nelson, 39–49. Albany: State University of New York Press, 1998.

Laidlaw, James. *Riches and Renunciation: Religion, Economy, and Society among the Jains.* Oxford: Clarendon Press; New York: Oxford University Press, 1995.

Lodrick, Deryck O. *Sacred Cows, Sacred Places: Origins and Survivals of Animal Homes in India.* Berkeley and Los Angeles: University of California Press, 1981.

Mahias, Marie-Claude. *Déliverance et convivialité: Le système culinaire des Jaina.* Paris: Éditions de la Maison des Sciences de l'Homme, 1985.

Mardia, K. V. *The Scientific Foundations of Jainism.* Delhi: Motilal Banarsidass, 1996.

Michaels, Axel. "La nature pour la Nature—Naturzerstorüng und Naturschonung im traditionellen Indien." *Asiatische Studient/Études Asiatiques* 50 (1996).

Modi, Resma. "Living a Jain Way of Life in the Western Environment." Youth essay contest entry, in Jaina Convention Souvenir, Pittsburgh, 1993.

Norman, K. R. "The Role of the Layman According to the Jain Canon." In *The Assembly of Listeners: Jains in Society,* edited by Michael Carrithers and Caroline Humphrey, 31–39. Cambridge: Cambridge University Press, 1991.

O'Connell, J. "Jain Contributions to Current Ethical Discourse." 1998 Roop Lal Jain Lecture. Toronto: Centre for South Asian Studies, University of Toronto.

Orr, Leslie C. "Jain and Hindu 'Religious Women' in Early Medieval Tamilnadu." In *Open Boundaries: Jain Communities and Cultures in Indian History,* edited by John E. Cort. Albany: State University of New York Press, 1998.

Pūjyapāda. *Sarvārthasiddhi (with* Tattvārtha-sūtra *of Umāsvāti).* Sanskrit text with Hindi translation by Phoolchandra Siddhāntaśāstrī. Jñānapīṭha Mūrtidevī Jaina Granthamālā, Sanskrit Grantha, no. 13. Varanasi: Bhāratīya Jñānapīṭha, 1971.

Reynell, Josephine. "Prestige, Honour and the Family: Laywomen's Religiosity amongst the Śvetāmbar Mūrtipūjak Jains in Jaipur." *Bulletin d'Études Indiennes* 5 (1987): 313–59.

———. "Women and Reproduction of the Jain Community." In *The Assembly of Listeners: Jains in Society*, edited by Michael Carrithers and Caroline Humphrey, 41–65. Cambridge: Cambridge University Press, 1991.

Sanghavi, Sukhal. *Pacifism and Jainism*. Banaras: Jain Cultural Research Society, 1950.

Schmidt, Hans-Peter. "Ahimsa and Rebirth." In *Inside the Texts, beyond the Texts: New Approaches to the Study of the Vedas*, edited by Michael Witzel. Cambridge, Mass.: Department of Sanskrit and Indian Studies, Harvard University, 1997.

Schweitzer, Albert. "Civilization and Ethics." In *The Philosophy of Civilization*, part 2. Translated by John Nash. London: Macmillan Publishing Co., 1929.

Seed, John. "Spirit of the Earth: A Battle-Weary Rainforest Activist Journeys to India to Renew His Soul." *Yoga Journal* 138 (1998): 69–71, 132–36.

Shah, Atul K. "Imprisoned by Success." *Young Jains International Newsletter* 11, no. 3 (1998): 5–6.

Shah, Pravin. "Jainism in the Next Millennium." In Philadelphia Jaina Convention Souvenir booklet, 1999.

Shanta, N. *La voie Jaina: Histoire, spiritualité, vie des ascètes pèlerines de l'Inde*. Paris: O.E.I.L., 1985.

Shiva, Vandana. *Staying Alive: Women, Ecology, and Development*. London: Zed Books, 1988.

Siddhasena. *Nyāyāvatara*, edited by A. N. Upadhye. Bombay: Jaina Sahitya Vikasa Mandala, 1971.

Singer, Peter. "All Animals Are Equal." *Philosophic Exchange* 1, no. 5 (1974): 243–57. Reprinted in *Environmental Philosophy: From Animal Rights to Radical Ecology*, edited by Michael E. Zimmerman et al., 26–40. 2d ed. Upper Saddle River, N.J.: Prentice Hall, 1998.

Singhvi, L. M. *The Jain Declaration on Nature*. London: Institute of Jainology, 1990.

Soni, Jayandra. "Philosophical Significance of the Jaina Theory of Manifoldness." In *Philosophie aus interkultureller Sicht*, edited by Notker Schneider, 283–85, Studien zur interkulturellen Philosophie, 7. Amsterdam and Atlanta, Ga.: Rodopi, 1997.

Sukhalaji, Pandit. *The World Pacifist Meeting and the Role of Jainism*. Ahmedabad: Publishing House, 1957.

Sūri, Śānti. *Jīva Vicāra Prakaraṇam, along with Pāthaka Ratnākara's Commentary*, edited by Muni Ratna-Prabha Vijaya. Translated by Jayant P. Thaker. Madras: Jain Mission Society, 1950.

Tobias, Michael. *Ahimsa: Non-Violence*. PBS Film, Los Angeles, Direct Cinema, 1989.

_______. "Jainism and Ecology." In *Worldviews and Ecology: Religion, Philosophy, and the Environment*, edited by Mary Evelyn Tucker and John A. Grim. Maryknoll, N.Y.: Orbis Books, 1994.

_______. *Life Force: The World of Jainism*. Berkeley: Asian Humanities Press, 1991.

———. *A Vision of Nature: Traces of the Original World*. Kent, Ohio: Kent State University Press, 1995.

_______. *World War III: Population and the Biosphere at the End of the Millennium*. Santa Fe, N.M.: Bear and Company, 1994.

Umāsvāti. *Tattvārtha Sūtra; That Which Is*. Translated by Nathmal Tatia. New York: HarperCollins, 1994.

Vallely, Anne. "Women and the Ascetic Ideal in Jainism." Ph.D. diss., University of Toronto, 1999.

Vyākhyāprajñapti Sūtra. 3 vols. Jaina Agama Series, no. 4, pts. 1–3, edited by Pandit Becaradasa Jivaraja Dosi. Bombay: Shrī Mahāvīra Jaina Vidyālaya, 1978–1982.

Vyākhyāprajñapti Sūtra. 4 vols. Jināgama Granthamālā, nos. 14, 18, 22, 25. Beawar, Rajasthan: Sri Agam Prakashan Samiti, 1982–1986.

Williams, R. *Jaina Yoga: A Survey of the Mediaeval Śrāvakācāras*. London: Oxford University Press, 1963.

Zwilling, Leonard, and Michael J. Sweet. "'Like a City Ablaze': The Third Sex and the Creation of Sexuality in Jain Religious Literature." *Journal of the History of Sexuality* 6 (1996): 359–84.

Notes on Contributors

Christopher Key Chapple is Professor of Theological Studies and Director of Asian and Pacific Studies at Loyola Marymount University in Los Angeles. He is the author of *Karma and Creativity* (State University of New York Press, 1986) and *Nonviolence to Animals, Earth, and Self in Asian Traditions* (State University of New York Press, 1993); co-translator of the *Yoga Sutras* (Sri Satguru Publications, 1990); and editor of several books, including *Ecological Prospects: Scientific, Religious, and Aesthetic Perspectives* (State University of New York Press, 1994) and, with Mary Evelyn Tucker, *Hinduism and Ecology: The Intersection of Earth, Sky, and Water* (Center for the Study of World Religions, Harvard Divinity School, 2000).

John E. Cort is Associate Professor of Religion at Denison University, where he teaches courses on religion in Asia and also on comparative themes (including courses entitled "Religion and Nature" and "Varieties of Environmentalism"). His research has focused primarily on the Jains. He is author of several dozen articles and of *Jains in the World: Religious Values and Ideology in India* (Oxford University Press, 2001).

Paul Dundas studied Sanskrit, Classics, and Middle Indo-Aryan philology at the Universities of Edinburgh and Cambridge and is currently Senior Lecturer in Sanskrit in the University of Edinburgh. He is the author of *The Jains* (Routledge, 1992).

Bhagchandra Jain received his D.Litt. in Sanskrit, Pali-Prakrit, and Hindi, and his thesis was on Jainism in Buddhist literature. A former professor and director of the Center for Jaina Studies at the University of Rajasthan, Jaipur, and Parshwanath Vidyapeeth Varanasi, he has been professor and head of the Department of Pali-Prakrit, Nagpur University. He is also Honorary Director of the Sanmati Research Institute of Indology, Nagpur, and Sarvodaya Jain Vidyapeeth, Sagar, Madhya Pradesh. He has written extensively on Jainism and Mahāvīra, Jain philosophy, Jain logic, and Buddhist philosophy, and has published approximately fifty books. He has received the Central Government of India Award, the National U.G.C. Fellowship, Commonwealth Fellowship, the Rampuria Award, and the Kundakundajnanapeeth Award, among others.

Padmanabh S. Jaini graduated from the University of Bombay and spent two years studying Theravada Buddhism in the Vidyodaya Monastery in Sri Lanka. He received his Ph.D. degree in Buddhist literature from the University of London. He has been on the faculties of the Banaras Hindu University and the School of Oriental and African Studies, London. He joined the faculty of the University of Michigan, Ann Arbor, in 1969 and moved to the University of California at Berkeley as Professor of Buddhist Studies in 1972. His major publications are in the field of Buddhist and Jain doctrines, among them *The Jaina Path of Purification* (University of California, 1979), *Apocryphal Birth-Stories* (The Pali Text Society, London, 1986), and *Gender and Salvation: Jaina Debates on the Spiritual Liberation of Women* (University of California, 1991).

John M. Koller, past president of the Society for Asian and Comparative Philosophy, is Professor of Asian and Comparative Philosophy at Rensselaer Polytechnic Institute. Author of *Asian Philosophies* (Prentice Hall, 2002), *The Indian Way* (Macmillan, 1982), *A Sourcebook in Asian Philosophy* (Macmillan, 1991), and numerous journal articles and book chapters, he has lectured at many universities in India and the United States and is an external Ph.D. examiner at four universities in India.

Satish Kumar is the director of programs at Schumacher College in Devon, editor of *Resurgence*, and founder of the Small School. At the early age of nine he became a Jain monk, and at eighteen he joined the Gandhian Movement. He later walked from India to the United States, covering 8,000 miles and propagating peace and nonviolence. His autobiography, *No Destination*, was published in the United Kingdom by Green Books and appeared in the United States under the title *Path without Destination: An Autobiography* (Eagle Brook, 1999).

Sadhvi Shilapi, a Jain nun, comes from Veerayatan, a Jain socioreligious institution located at Rajgir in the State of Bihar in northeast India. She received her master's degree in comparative Indian religions from King's College, London, in 1995 and is currently working there on her doctoral thesis, entitled "Compassionate Aspect of Jainism." She has established an International School of Jainism with centers in North and South London and in Nairobi and Thika in Kenya.

L. M. Singhvi has been a member of the Indian Parliament (Lok Sabha) from 1962 to 1967; president of the Supreme Court Bar; chair of the Human Rights Congress (1990); high commissioner for India with cabinet rank in the U.K. (1991–1997); member of Parliament (Rajya Sabha, 1998–2004); chair, High Level Committee on Indian Diaspora with cabinet rank; and one of the presidents of the Council of a World Parliament of Religions.

Kim R. Skoog is a member of the philosophy program at the University of Guam and currently is Chair of the Division of Humanistic Studies (a multidisciplined unit focusing on the human prospect). He received his Ph.D. from the University of Hawaii in Comparative Philosophy, his M.A. from the University of Washington, and his B.A. from the University of Minnesota. He has presented and published numerous articles both in Asia and the United States. His most recent publication is a chapter in *Living Liberation in Hindu Thought*, edited by Andrew O. Fort and Patricia Y. Mumme (State University of New York Press, 1996).

Nathmal Tatia was research director of the Jain Vishva Bharati Institute in Ladnun, Rajasthan, until his death in 1998. He received his B.A., M.A., and Ph.D. degrees from Calcutta University where his doctoral thesis was on Jain philosophy. He taught in Nalanda and in Vaishali before going to Ladnun in 1977. He was a visiting professor of Buddhist studies at Harvard University in 1990 and attended the International Congress of History of Religions at Claremont in 1965 as the representative from India. He published numerous books, articles, translations, and edited works, including a bibliography of Jain research (*Tulsi Prajna*).

Anne Vallely teaches at the Universities of Concordia and McGill in the fields of anthropology, religion, and environmental ethics. Her Ph.D. degree in anthropology from the University of Toronto was based on ethnographic research among a Jain mendicant community in Rajasthan. She is the author of *Guardians of the Transcendent: An Ethnography of a Jain Ascetic Community* (university of Toronto Press, 2002).

Kristi L. Wiley is a visiting lecturer at the University of California at Berkeley where she teaches courses on religion in South Asia, including a course on religion and ecology. Her research has focused on Jainism, primarily on karma theory. She is the author of the *Historical Dictionary of Jainism* (Historical Dictionaries of Religions, Philosophies, and Movements Series; Scarecrow Press, forthcoming).

Index

mithyātva, 35
mokṣa
household activities impeding path to, 51
human capacity for, 45, 142
mokṣa-mārga ideology, 70–71, 73, 90n.24
as not progressive and linear, 40
in orthodox understanding of *ahiṃsā*, 212
as possible during certain periods of time, 58n.50
soul's separate identity in, 49
monasticism (asceticism)
avoiding harm in, xxxiv, 142
begging, 100, 114n.16, 182, 198–99
as conscience of Jain tradition, xxxviii
daily life of Indian ascetics, 197–99
as decreasing in significance among North American Jains, 203–4, 206, 211
elephant ascetics criticizing, 100
as exemplifying Jain values, 142
handbooks on lay behavior written by monks, 104
in Jain hierarchy of existence, 202
laypeople supporting, 102, 143, 198
lifestyle as environmentally friendly, 137–38
mahāvratas, 8, 46, 71, 74
meditation in, 83
migration in, 82
as not involved in damage to environment, 8
as not taking root in Jain diaspora, xlii, 195
number of monks and nuns, 142–43
offerings to mendicants, 146
polarity of laypeople and, 143
practice based solely on renunciation, 71
pratikramaṇa performed in, 75
universal observance of Jain principles by, 71
women as mendicants, 77, 148
women providing food to mendicants, 148, 198
monks. *See* monasticism
morality. *See* ethics
mosquitoes, malaria control through killing, 152
mountains, 182–83
Mṛgapūtra, 31–32
Muhammad Tughlak, 12
Muir, John, 135
Mūlācāra, 174, 176
Mushara, 165
myth, transformative power of, 68–69

Nabhan, Gary, 69
naigama, 25, 26–27
nāma karma, 40–41, 55n.24
Nandy, Ashis, 91n.32
nāraki (hell beings; infernal beings), 36, 39, 47, 91n.36
narrative, transformative power of, 69
natural disasters, Jain aid to victims of, 147
natural resources. *See* resources
nature
autonomous value lacking in Jain worldview, xxxviii–xxxix, 96
balancing rights of humans and of, 79
causes of rift between humans and, 131–32
consequences of human mastery of, 170
culture contrasted with, 201
dualism in Jain view of, xxxviii–xxxix, 70, 73, 95–96
gender difference and, 76–77
in human degradation, 97–98
humanity's place in Jain worldview, 3–4
Indic words for, 67
Jain perspectives on, 35–59
Jain theories of, 1–59
learning from, 181
living in harmony with, 184
as moral theater, 202, 213
mountains, 182–83